MONKEYS WITHOUT TAILS

JOHN NAPIER

TAPLINGER PUBLISHING COMPANY
NEW YORK

First published in the United States in 1976 by
TAPLINGER PUBLISHING CO., INC.
New York, New York

Library of Congress Catalog Card Number: 75-33425
ISBN 0-8008-5322-9

Contents

Acknowledgement is due to the following for permission to reproduce illustrations; the numbers refer to pages. *Black and white*, 8, S. L. Washburn; 10, (top) Jen and Des Bartlett, Bruce Coleman Ltd, (bottom) Jerry Cranham for BBC; 19, (Photograph) C. K. Brain, (drawings) Maurice Wilson; 29, D. M. Sorby; 30, Popperfoto; 39, Werner Curth, Ardea; 40, (top) J. Allan Cash, (bottom) M. Kawsi; 41, M. Kawsi; 43, (top & centre) Zoological Society of London, (bottom) Fox photos; 45, (top) Fox photos, (centre & bottom) Zoological Society of London; 50, St Thomas's Hospital Medical School; 51, (top & bottom) Claire Leimbach; 57, From 'The Origins of Man' Pub. The Bodley Head; 61, Trustees of the British Museum (Natural History); 65, Blackwell Scientific Publications Ltd; 67, Trustees of the British Museum (Natural History); 71, Smithsonian Institute; 77, Kobal Collection; 81, By permission of Baron Hugo van Lawick, In the Shadow of Man, Pub. Collins; 93, (left and centre) By courtesy of R. J. Harrison and W. Montagna; 101, P. R. Davis; 102, (top) Press Association, (bottom) Popperfoto; 107, (top left) Popperfoto; 113, BBC; 116, RTHPL. *Colour, Zoological Society of London*, Ring tailed lemur, Bushbaby Galago, Slender loris, Cottontop tamarind, Squirrel Monkey, Colobus, Diana monkey, Pigtailed macaque, mandrill; *R. D. Martin, Pic on Tour*, Mouse lemur; *Fox photos*, Potto; *David Attenborough*, Tarsius, Female in characteristic hanging position with young infant; *Shin Yoshino*, Douroucouli, Common marmoset, Red-faced Uakari, Spider monkey, Proboscis monkey, Langur, Debrazza's monkey, Hot-spring bathing, Japanese macaques from Northern Honshu. Riding out the snowy winter in spite of food in short supply, chimpanzee, Gorilla; *Keystone*, White-fronted capuchin; *Des & Jen Bartlett, Bruce Coleman Ltd*, Patas monkey; *S. L. Washburn*, Baboon; *Sasson—Robert Harding*, Open savanna, East Africa; *Claire Leimbach*, Rain forest, Borneo, Adult male Bornean orang-utan; *Sunday Times*, male motorist; *National Geographic Society*, Chimp with a tool, Chimpanzees of Gombe Stream National Park and Jan Van Lawick-Goodall.

The drawings are by Richard Bonson.

Acknowledgements

My primary acknowledgement is to the Royal Institution whose founding fathers had the wisdom to see then what we are just beginning to realise now – the importance of presenting, in digestible form, science for the citizen. The first Christmas lecture for young people was given in 1826, and between 1827 and 1860 the great Michael Faraday gave these lectures on nineteen occasions. Faraday illustrated his discourses with many elegant demonstrations, thereby setting a tradition which is followed to this day. The lecture series on which this book is based was the 141st of the genre.

The formidable responsibility for presenting these lectures with one eye on tradition, one on the interests of the contemporary audience, and one (a third, or pineal eye is an essential requisite for the lecturer) on the demands of television presentation, would have been impossible without the wonderful help, perceptiveness and understanding of the RI and BBC personnel involved.

To Professor Sir George Porter, the Director of the Royal Institution, to Professor King and to Mr Bill Coates and his assistants who built and operated most of the demonstrations, I record my most appreciative thanks and unbounded admiration. To Alan Sleath of the BBC, the producer of the series, goes my fullest respect for his technical skill, but also, more importantly, my gratitude for his wonderful kindness before, during and after that marathon fortnight.

Special thanks are due to Mollie Badham and Natalie Evans, proprietors of Twycross Zoo Park, who in spite of snowdrifts several feet deep on the A444 managed to get their chimpanzee Judy to 'the church on time'. Judy, as those who saw the programmes will remember, was the real star of the series. I am also most grateful for the co-operation shown by the Director and Trustees of the British Museum (Natural History) in making many of their specimens available to me including the neck and head of Geronimo, the giraffe. I can now reveal that my claim in the first lecture that we had had to hollow out the floor to accommodate Geronimo in his entirety was false – he never was anything but a beautiful neck!

My thanks are also due to several people who have helped turn six rather rambling lectures into what, with due modesty, I think is a passable book; to my wife Prue, whose role was varied – dresser, chauffeur, groomer and morale-booster during the lectures, and principal critic and researcher during the writing; to Nicholas Lumsden who was responsible for translating the manuscript from the original giraffe and for typing the final product; and to Dr Gilbert Manley who has seen to it that my zoological statements fall within the bounds of reason.

Finally, I am indebted to the many scientists whose knowledge and opinions are quoted herein without, in many cases, specific acknowledgement. I have borrowed their wisdom. I can only hope that when the 200th Royal Institution Christmas Lectures are presented in the year 2032 someone will borrow some of mine.

J.N.

Preface
A giraffe's-eye view
of man and apes

Anatomically and behaviourally man sees himself as a unique species, as the President of the living world and the principal source of its power and its glory. But is this the true picture of man's place in nature? Compared with ancient life forms such as the jawless fishes, for instance, whose history extends back in time for 350 million years, modern man – *Homo sapiens* – is the merest newcomer. As a species, man is in his infancy. His impact on the world is undeniable but the world could get on very well without him, the seasons would come round just the same and the tides would still ebb and flow.

To see ourselves as we really are and not as we imagine ourselves to be, we need the viewpoint of a non-human non-primate observer, and so in these pages we shall be looking at man through the eyes of a contemporary mammal, the giraffe. The giraffe is tall and has an excellent range of vision; it lives in Africa among several species of contemporary primates and has a fossil history that, being as long as man's, gives it the right to speak authoritatively about the stages of human evolution that have been going on under (albeit a long way under) its nose. A giraffe sees man as a monkey without a tail, a rather ordinary primate that differs very little from the baboons, patas monkeys and savanna monkeys that are its daily companions.

Naturally the views and opinions on which this book is based have been translated from the original giraffe and have, no doubt, lost something of their original freshness of approach.

1 Monkeys *with* tails. Baboons, giraffes' daily companions.

1

A walk round the primate zoo

Monkeys without tails are men but they are also apes. However, men and apes share more than taillessness as I hope this book will demonstrate. There are four apes: the gibbon, the orang-utan, the gorilla and the chimpanzee. The gibbon is often called a Lesser Ape and the other three, Greater Apes. Of these it is the chimpanzee that is most familiar to most people and, perhaps more important, is the best understood scientifically. So for the purposes of this book the chimpanzee is appointed to represent the ape branch of the monkeys without tails.

The chimpanzee is not a bad choice because it is a very easy animal for humans to identify with. Chimpanzees are romping extroverts (a rather rare characteristic in mammals); their faces are full of expression which man – rightly or wrongly – interprets in terms of his own experience and, when young at any rate, they emanate a sensitivity, a warmth of feeling, an intelligence and an acuteness of understanding that persuades man that chimps are more deserving of human patronage than most 'dumb' animals. Such benevolence has not always character-ised the human attitude towards apes. Before Darwin made them respectable with his evolutionary theories, they were the whipping boys of cartoonists and churchmen who saw them as scapegoats for all that was vile and libidinous in man. Even today we are not totally free of this calumny, and while the Church has found other scapegoats, the cartoonists still use the ape as the symbol of rapacity and aggression.

Even the most understanding of us still experience a sense of revulsion in the face of the innocent sexuality of the apes while illogically we condone the most overt demon-strations of not-so-innocent sexuality in our own species.

The title of this book needs some defend-ing. On the face of it to refer to men and apes as monkeys is a major zoological gaffe; in academic circles men are men, apes are apes, monkeys are monkeys and lemurs are lemurs, but in common parlance all non-human primates are monkeys – and monkeys have tails. This is a splendid piece of rationalisation on my part to justify a title of which I am particularly fond, but were it not for the second and better reason I wouldn't have a leg to stand on. The second reason is that in this book we are looking at man and apes through the eyes of a giraffe. Giraffes are in daily contact with baboons, vervet monkeys and patas monkeys all of which are well endowed with tails. Although chimpan-zees and humans are less familiar features of a giraffe's daily round one can feel sure that in spite of human trappings like clothes, spears, guns and Land Rovers, giraffes are not blind to the essential primate-ness of man.

From a giraffe's elevated point of view, eighteen feet more or less above the gound, it is enough to distinguish two classes of primate: those with tails and those without. As classifications go this is not bad inasmuch as it distinguishes, firstly, the apes from the monkeys and, secondly, links the apes with man; but like all biological rules, however good, there are a number of exceptions.

In general it is true to say that all true monkeys possess tails. Monkeys as a whole are arboreal creatures and in the world of trees the tail has two major functions: it acts as a balancing device (just as a pole provides side-to-side stability for a tightrope walker) and it serves as an air-brake (as flaps do in an aeroplane). There are a number of other uses (including communication) to which the tail is put but these are secondary to its basic locomotor role. The exceptions to the rule are found among some, but not all, ground-living primates. The macaque monkeys of the Celebes and of the Atlas Mountains of North Africa, for example, are tailless; others like the macaques of Japan possess only the merest apology for a tail. Even scientists refer to the Celebes macaques and the Barbary macaques of the Atlas Mountains of North Africa as 'apes'. This

2
Monkeys
without tails.
Young
chimpanzee
and a
young man
swapping
roles.

attitude accords with the well-established giraffid view that the only non-human primates which lack a tail are apes and men. Baboons, the most ground-living of all primates except man, oddly enough retain a very respectable tail. They provide an example of the secondary function of tails mentioned above. Most species of baboon do in fact climb trees for food and they also sleep in trees, but the tail is retained even in those species, like the arid-country Hamadryas or sacred baboons, which never see a tree, let alone climb it. The most likely explanation for the retention of a tail in baboons is that it provides a first-class backrest for the infant which for the second six months of its life rides on its mother's rump, jockey-style. The sharp upcurve of the baboon tail from its root provides a convenient and effective orthopaedic support. This may seem a trivial reason for retaining an apparently useless character but it is one of a countless number of examples of atavism in animals. Why does man retain his apparently useless but troublesome appendix, or that ridiculous feature, the lobe of the ear?

The giraffe's simple classification – monkeys with and monkeys without – does not, however, go quite far enough for our purposes. There are approximately 186 different species of primates and about 523 subspecies or races, so to find one's way about such an assemblage, some system of indexing has to be devised. The art of indexing whether in an office or a museum is to arrange things in logical groups. The question of what is or what is not logical is the essence of both business practice and zoological procedure.

The principles of modern zoological classification were laid down by the Swedish naturalist Carl Linnaeus (1707–78) and to Linnaeus it seemed logical to arrange animals in certain groups because they looked alike or behaved alike. We know a little better now and we are aware that although the similarity of animals often means that they share a common ancestry, this is not always the case. For example, both porcupines and hedgehogs have spines on their back to protect them against predators but this is as far as their similarity goes. Hedgehogs belong to the Order Insectivora and porcupines to the Order Rodentia. Many of the Australian mammals, the marsupials, have physical counterparts in the rest of the world; there are marsupial 'moles', 'cats', 'dogs', 'bears' and so on. The explanation of this evolutionary mechanism, known as convergence, is really quite simple: given broadly the same altitude, rainfall and temperature conditions, animal environments around the world are extraordinarily similar. Each biome, as it is called, imposes certain physical, physiological and behavioural demands on the animals that occupy it; if animals can live up to these conditions they will survive, if not they will die and the species to which they belong will eventually become extinct. This is the basis of the theory of natural selection discussed more fully in chapter 2.

Tropical rain forests in south-east Asia differ only in detail from the rain forests of South America and Africa. High-altitude, rocky, treeless moorland is much the same in Ethiopia as it is in Chile or Peru. Coniferous forests of Canada have their counterparts in other high-latitude areas of the northern hemisphere. Each geographical realm has produced its own animal answers to a universal challenge and it is not surprising that the end-products should be so similar.

Nowadays our main criterion for classifying animals is based not simply on *physical* similarity but on *genetic* similarity. Genes are the units of heredity which dictate how an animal looks and to some extent how it behaves. Most animals that look alike are genetically related but the likeness must be homologous (identical) and not just analogous (similar). The fundamental unit of any classification is the *species* because a species is a real thing which can be precisely defined and observed in nature; the species is based on genetic affinity because – by definition – two animals that interbreed and produce viable and fertile offspring belong to the same species. Only animals very closely related genetically are capable of interbreeding. The so-called higher categories of

3 In animal classification higher categories 'contain' lower categories like a set of Russian dolls.

classification within a single order – genus, family, superfamily, infraorder, suborder and so on – are not 'real' in the same sense as are species, they represent a purely arbitrary means whereby like objects can be gathered together in convenient categories, the largest embracing the next largest and so on, like a set of Russian dolls (Figure 3).

Although hardly any two primatologists agree in detail about the classification of primates, there is a general consensus of opinion on major issues. The place of the tarsier has always been contentious, and its appearance as a separate suborder in my scheme will no doubt make a few colleagues bristle.

There are three major groups (suborders) of primates: the lemurs and lorises (prosimians), the tarsier and the anthropoids, which include the apes, monkeys and man.

Although prosimians and anthropoids shared a common ancestor in the beginning (and therefore looked alike), the differences between them today are profound. Natural selection operating through the environment has produced forms that look and behave in a quite distinctive fashion, so much so that some scientists have argued the case for the separate ordinal status of lemurs. However, there is little real justification for such a viewpoint. The more we learn about lemurs and lorises, the more primate-like we find them to be.

CLASSIFICATION OF THE PRIMATES

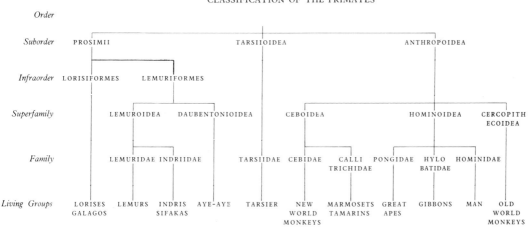

THE PRIMATE ZOO

There are many splendid zoos in Britain, Europe and the USA where a representative range of primates can be seen. I plan to lead you on a conducted tour around an imaginary zoo which in my model, for simplicity's sake, consists of a single building. The inside cages are shown but not the outside runs; I emphasise this point because access to the sun, rain, warmth and cold as well as to an area in which something like the normal range of locomotor activity can be indulged in, are essential ingredients for good primate welfare and husbandry. The descriptions of the twenty-one commonest species are planned to bring out the important characters and characteristics that an observer can easily see under zoo conditions. I have paid less attention to teeth, for example, than to hands and feet.

The old-fashioned zoo was based on what is known as the Noah's Ark Principle and primates were exhibited in pairs. Early zoo-keepers did not know that only a very few species of primates actually enjoy this type of domestic bliss in the wild. In fact most species of primates live in social groups consisting of large numbers of males and females of all ages. The Noah's Ark Principle is gradually being replaced by the Natural Group System.

People often ask just how much normal behaviour can be seen in captive animals in zoos. The answer is: much more than one might suppose. In a recent study of the Hamadryas baboon in its natural environment in Ethiopia and in its artificial environment in Zürich Zoo, Dr Hans Kummer reports that fifty-nine patterns of behaviour seen in the wild were also seen in the splendid, roomy outdoor enclosure of the Hamadryas colony at Zürich. To be fair, Kummer also reported that an additional nine behavioural patterns seen in the Zoo were not seen in the wild. This, then, is a measure of the effects of an abnormal environment on a wild animal; however these 'vices' as they are termed (without necessarily having any vicious implications) are also seen in farm livestock kept under intensive farming conditions. Under less satisfactory zoo conditions the proportion of abnormal to normal behavioural patterns rises rapidly. Good zoo-keeping – and good farm livestock management for that matter – can be judged by the *behaviour* of animals.

Our tour starts with the Prosimians and the Tarsiers, two of the three primary subdivisions of the Order Primates. Prosimians as a group are nocturnal in habit. There are a few exceptions, the commonest and best known of which is the ringtailed lemur (*Lemur catta*). So before passing through the baffle screens into the Nocturnal House – the moonlight world of creatures whose day is our night, we pause at the cage containing the elegant and appealing ringtails.

Cage 1 *Ringtailed lemur* (*Lemur catta*). Madagascar. The ringtail is the most ground-living of all prosimians and lives in social groups of about sixteen individuals. Their diet is wholly vegetarian. Their overall colour is a light grey, their faces show a black mask which encircles the muzzle and the eyes; their tails are white, banded with black. There is no great difference in size between males and females, but the male can usually be distinguished by a prominent naked black patch on the middle of the inner surface of the forearm. This is a special gland concerned with the ringtail's characteristic behaviour of scent marking, an important method of communicating information to others of its species; another glandular area lies in the region of the genitalia, and ringtails can often be seen 'marking' by rubbing this area against the posts and poles in their cages for purposes not fully understood but possibly related to defining their territory.

NOCTURNAL HOUSE

Cage 2 *Mouse lemur* (*Microcebus murinus*). Madagascar. Among the smallest of all primates, the mouse lemur does not really live up to one's idea of what a primate should look like; instead its appearance is more in keeping with its common name. Mouse lemurs are strictly nocturnal and probably solitary animals in nature and vegetarian and insectivorous in diet. They are up to five

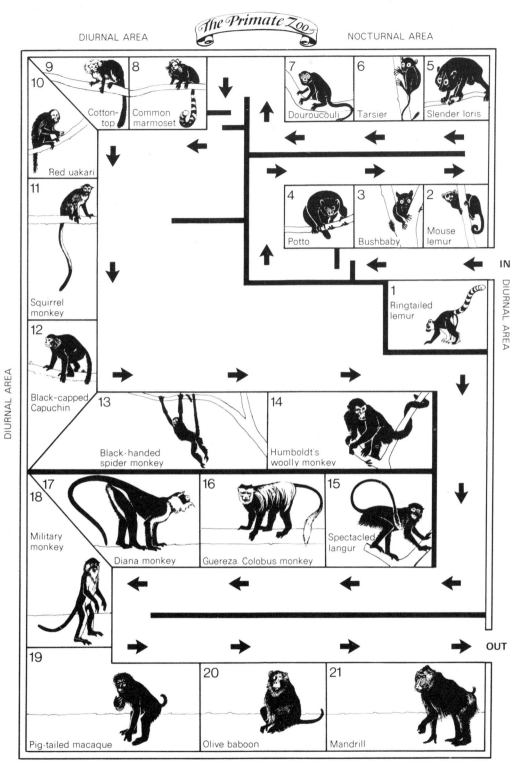

The Primate Zoo

DIURNAL AREA NOCTURNAL AREA

9 Cotton-top
8 Common marmoset
10 Red uakari
11 Squirrel monkey
12 Black-capped Capuchin

7 Douroucouli
6 Tarsier
5 Slender loris
4 Potto
3 Bushbaby
2 Mouse lemur
1 Ringtailed lemur

IN

DIURNAL AREA

13 Black-handed spider monkey
14 Humboldt's woolly monkey
17 18 Military monkey
Diana monkey
16 Guereza. Colobus monkey
15 Spectacled langur

OUT

19 Pig-tailed macaque
20 Olive baboon
21 Mandrill

DIURNAL AREA

4 Prosimians, New and Old World monkeys in a stylised 'monkey house'. Colour photographs of these species are shown following page 16. The apes are not illustrated here.

inches in length and have short, plushy grey-brown fur. The hair forms dark rings around the eyes and a white streak down the nose which is short and pointed. One peculiar feature of their physiology is that mouse lemurs store fat at the roots of their tails, a device that permits them to employ a modified form of hibernation called torpidity which lasts for several weeks during the Madagascan winter.

Cage 3 *Bushbaby* (*Galago senegalensis*). Widespread over Africa. This species is one of the smallest of the galago branch of the prosimians and weighs 200-300 grammes; males are slightly heavier than females. The general appearance of bushbabies – the pointed muzzle, large round eyes and the large transparent and very mobile ears – is reminiscent of mouse lemurs. However, there is one very striking difference: the bushbaby has extremely long hindlimbs (compared with forelimbs) which act as powerful springs. The prodigious leaps of bushbabies in an almost vertical direction from a standing start can be as much as fifteen times their body length.

Like all primates, galagos have hands rather than paws; their thumbs and big toes are well separated from the rest. The nails are flat, not claw-like, and the fingertips are broad and sensitive. The second toe bears a special claw which is used for grooming the fur and scratching. Galagos live in forests and woodlands over wide areas of Africa; their diet is largely composed of insects but some fruit is also eaten.

Bushbabies make attractive pets but are not as gentle and cuddly as their name might suggest; furthermore, being nocturnal, their period of greatest activity is the opposite of man's. To enjoy your pet you must be prepared to sit up late at night.

Cage 4 *Potto* (*Perodicticus potto*). West and central Africa. While the galagos are exceptionally quick-moving, pottos are sloth-like in their slowness. Watching the movements of the potto is like seeing a film in slow motion; each branch is grasped deliberately with broad pincer-like hands and feet and then, just as deliberately, released so that the potto seems to flow along the branch. In order to avoid obstacles the potto will follow a spiral course, transferring from time to time to the under surface of the branch. This stealthy progress which does not disturb the twigs or leaves is probably an adaptation for stalking prey – insects, lizards and nestling birds.

Pottos have a plushy brown coat and are stockily built. They are quite incapable of leaping like galagos. Their faces are short and broad and their ears small and close-set; their tails are so small as to be ridiculous. One anatomical curiosity which has excited a lot of speculation is the presence of a row of five or six skin-covered vertebral spines on the back of the neck concealed in the fur; they may act either as offensive or defensive weapons, nobody is quite sure.

Cage 5 *Slender loris* (*Loris tardigradus*). Ceylon and South India. In terms of slowness and reduction of the tail, the slender loris is like a potto but that is where the resemblance ends. While the potto is plump and stocky, the loris is grotesquely thin. Like the mouse lemur and the bushbaby, the loris has a white streak down its nose and dark rings round its eyes. Like the galago and the potto, the loris is an insect-eater but it does not turn up its sharp-pointed nose at lizards and nestling birds. The slender loris shares a curious scent-marking habit with the galago; they may either secrete a few drops of urine as they creep along a branch or they may wash their feet and hands in a flow of urine which is then transferred to the branch at each step.

Cage 6 *Tarsier* (*Tarsius bancanus*). Southeast Asia. The tarsier is not a prosimian, neither is it an anthropoid; it occupies an intermediate position and shows many of the characteristics of both these major divisions of the primates. One of the characteristics it shares with prosimians is solitariness; it is also a creature of the night and an insect-eater. Tarsiers also have the very long hindlimbs of the galago and, as a con-

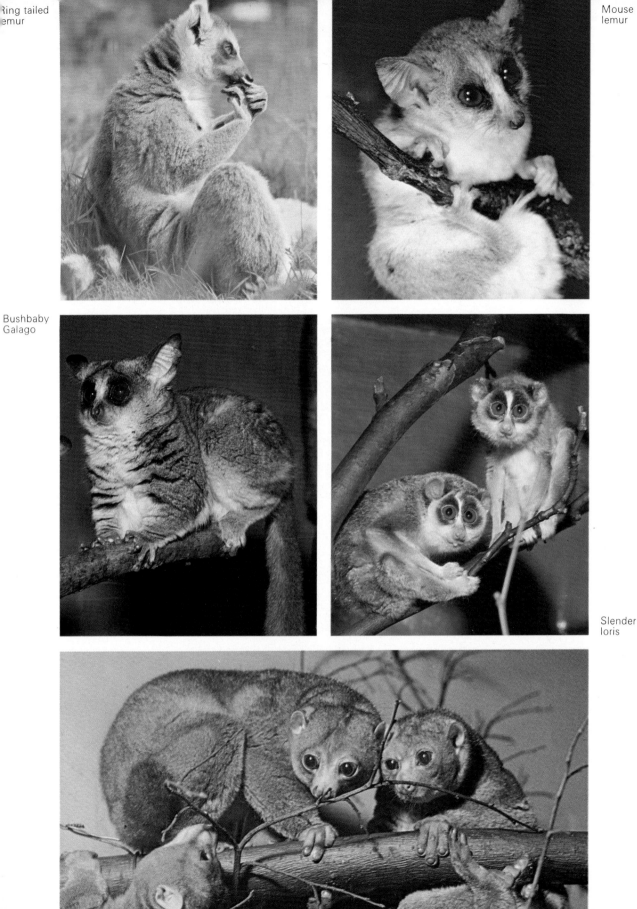

Ring tailed lemur

Mouse lemur

Bushbaby Galago

Slender loris

Potto

Tarsius

Dourouco

Common
marmoset

Cottontop
tamarind

Red uakar

Squirrel
monkey

White-fronted
capuchin

Spider
monkey

Proboscis
monkey

Langur

Colobus

Diana
monkey

Debrazza's
monkey

Patas
monkey

Pigtailed
macaque

Baboon

Mandrill

sequence, employ a similar type of leaping locomotion. Dissimilar characters are mainly anatomical; for example tarsiers have the 'dry' nose and mobile upper lips of monkeys and apes (as opposed to the 'wet' nose of lemurs and lorises), a dental pattern that seems to owe more to the anthropoids than to the prosimians, a placenta of anthropoid and not prosimian structure, and, finally, a brain that is relatively larger and more complicated than that of prosimians.

The tarsier has certain unique physical characters that appear to have no parallels in either of the other two groups, the Prosimians or the Anthropoids. For instance, the tarsier can turn its head through 180 degrees so that it is looking directly backwards; in this manner the forward set of its eyes is compensated for by a possible 360-degree scanning range. Tarsiers possess the unique primate adaptation of expanded friction pads on the tips of the fingers and toes; they act rather like the suction pads on a tree-frog's digits. Finally, tarsiers possess a unique type of thumb opposability.

Tarsiers are of peculiar interest to man because they represent an intermediate grade between prosimians and anthropoids which at least one authority in the past has considered to be antecedent to the human line. There is no real evidence for this belief, however. Tarsiers are the sole living representatives of a once widely dispersed suborder. They have changed little over forty-five million years and have been described as living fossils.

Cage 7 *Douroucouli (Aotus trivirgatus).* Central and South America. The douroucouli is a monkey – an anthropoid – and as such is rather out of place in the Nocturnal House which is the territory of the prosimians and tarsiers. The explanation is simple, the douroucouli is the only nocturnal monkey in the world. Douroucoulis are small creatures, about small-cat size, and with their large round eyes and pointed nose they have a prosimian 'look' about them. The facial similarity is an example of the evolutionary mechanism of convergence discussed above. In every other way they are good anthro-

poids. Their fur colour, as in many primates, is an undistinguished greyish-brown. The special markings of their coat are three dark lines on the head and white semilunar patches above the eyes in the eyebrow region. When douroucoulis are asleep these patches look for all the world like a pair of wide-awake eyes; this adaptation presumably provides a defence mechanism against daytime predators.

Douroucoulis live in family groups and like several other South American monkey species the father, not the mother, carries the young on his back.

We are now leaving the Nocturnal House, and we exit through a baffle-like arrangement of partitions which, while keeping the daylight out, permits free passage for visitors on their return to the ordinary world where day is day and night is night.

NEW WORLD MONKEY HOUSE

Cage 8 *Common marmoset (Callithrix jacchus).* Brazil, south of River Amazon. Marmosets are tiny monkeys that differ from all others in possessing claws on their hands and feet in the place of nails. Only the big toe has a typical flat primate nail. They are tree-livers and move along the branches with rapid squirrel-like movements using their claws to grip the bark; they seldom leap or drop as other monkeys do; even very small branches provide them with support. For food they depend on insects as well as fruits, flowers, buds and leaves. Marmosets feed by hand, plucking the food and bringing it to their mouths as squirrels do but, unlike squirrels, they can feed by using one hand only, not two; this behavioural pattern indicates that in spite of possessing claws, their hands are prehensile – a primate characteristic. Their social life is based essentially on the family unit; in fact they are believed to be monogamous. Twin births are quite usual but, like the douroucouli, it is the father who carries the infants, only handing them over to the mother for breast feeding. This behavioural pattern is a useful adaptation of small primates in which twin births are usual.

There are at least three species of marmoset and they differ principally in their 'adornments', particularly around the ears which may be bare and crimson (*C. argentata*), with white fan-like tufts (*C. humeralifer*), or, in the case of the common marmoset (*C. jacchus*) decorated by black, white, or yellow plumes of hair. The tail whatever its colour is often ringed.

Cage 9 *Cotton-top tamarin (or pinché)* (*Sàguinus oedipus*). Central and South America. The overall appearance of pinchés or tamarins is similar to marmosets but they are a little bigger and the tail is never ringed. They show a variety of adornments on face and body. Cotton-tops have tiny white hairs superimposed on a black face and a striking crest of long white hairs on the top of the head. Tamarins – like marmosets – are bright, vivacious little creatures with the air of immense curiosity and intelligence. In captivity they follow the movements of the observer with bird-like tilts of the head which at times seems to turn through 180 degrees in the vertical plane.

Cage 10 *Red uakari (Cacajao rubicundus)*. South America, upper Amazon. You will probably experience a severe shock on seeing this primate for the first time. The red uakari is an amazing-looking animal with a bald, crimson face, a doleful expression, a shaggy rust-coloured coat and a bobbed tail. In spite of a most unathletic appearance, uakaris are extremely acrobatic and agile in their native arboreal habitat. Very little is known about this primate in the wild but they are believed to live in moderately large social groups. Uakaris are not easy to keep in captivity but specimens can be seen at Twycross Zoo in the UK, at Frankfurt Zoo in West Germany and San Diego Zoo in the USA. There are two other species but they have rarely been exhibited in the western hemisphere.

Cage 11 *Squirrel monkey (Saimiri sciureus)* Widely distributed in Central and South America. A small and very appealing monkey – unfortunately too appealing for its own good. It is extensively used in medical research in the USA but also cruelly exploited by the pet trade, and by street photographers in London.

Squirrel monkeys are neat and elegant primates with short dense fur of various shades of rust or grey with yellow arms and legs. The head is oval in profile, the muzzle is short and black. The eyes are surrounded by broad white rings and the head hair which is dark is produced into a widow's peak dipping between the eyes. They are extremely active in the trees aided by a partly prehensile tail; being small, they have a wide feeding range in trees; they eat insects and fruit and live in large groups of up to 200 animals.

Cage 12 *Black-capped capuchin (Cebus apella)*. Central and South America up to 7000 feet altitude. Capuchins are the best known of all New World monkeys simply because, being tough and hardy animals which will eat virtually anything, they take well to captivity. The capuchin is the classic 'organ-grinder's monkey' which, dressed in a jacket with a fez on its head, will go through its paces, and survive, under the most unsuitable conditions. Capuchins come in a number of different shades of fawn through brown to black. *C. apella* has a dark chocolate-coloured fur but *C. albifrons* (white-fronted capuchin), also commonly seen in captivity, is cinnamon-coloured. Black-capped capuchins have black hair on their heads, which is elevated in symmetrical tufts or 'horns'.

In terms of locomotion, capuchins are very much all-purpose animals that can run, leap and climb. Their tails, which are partly prehensile, are usually held with the tip coiled; when capuchins are sitting on a branch the tail will wrap around the most convenient anchorage in a spiral fashion.

Their hands are extremely dextrous. The thumb acts independently of the fingers; however the degree of freedom and mobility of the thumb is in no way comparable to that of the Old World Monkeys, inasmuch as they lack the crucial element of true opposability of the thumb (see chapter 5).

5 Some external characteristics of Old and New World monkeys. Hard-skinned 'sitting pads' in an Old World monkey which are absent in New World forms and comparison of external noses. A. Old World monkey; B. New World monkey.

A

B

Cage 13 *Black-handed spider monkey* (*Ateles geoffroyi*). Central America. Spider monkeys are the greatest specialists of all the South American monkeys. In addition to all the basic New World monkey characters, spider monkeys possess a prehensile tail which is in constant use as a third hand; anything the hand can do, the prehensile tail can do equally well.

The similarity of tail to hand goes beyond simple grasping; it is also sensitive. On its underface the last third of the tail is hairless and arranged in a series of ridges which provide both a friction surface and an acutely sensitive tactile area, identical in form and function with the terminal pads of the finger tips. Spider monkeys can, and frequently do, suspend themselves by the tail alone. They move about in the trees either quadrupedally or by suspending themselves and swinging using various combinations of arms, legs and tail. Another specialisation, not easily explained, is the absence of the thumb.

Spider monkeys are quite large animals; they are about the size of a greyhound. The body fur of the black-handed spider monkey may be either gold, red, buff or dark brown, while the hands, feet, crown of the head and the knees are black.

Cage 14 *Humboldt's woolly monkey* (*Lagotrix lagothricha*). Widely distributed in South America. Woolly monkeys are a chubby version of spider monkeys; they share the blessings of a prehensile tail but they differ from spiders in the possession of a thumb. Woollys have short, dense, plushy fur. The face is short and although the nostrils are widely separated, as in all New World monkeys, the face has a rather 'human' look. Temperamentally, woollys are comparatively tractable creatures and can become very devoted to their keepers and owners but, like all adult primates, they are unreliable and totally unsuitable as pets.

We move now from the New World to the Old World monkeys. The differences between them are not very obvious. Overall, Old World forms are rather bigger and rather more highly coloured. Socially they are more advanced and physically and mentally they are definitely a step up the primate ladder. The two simplest of the rules-of-thumb that can be used to distinguish Old and New World forms are the shape of the nose (Figure 5) and the appearance of the bottom (Figure 5). Thus New and Old World monkeys can be distinguished coming or going.

The Old World monkeys are sharply divided into two major ecological groups, recognised taxonomically by placing them in two subfamilies:

Colobinae. *Langurs, colobus monkeys, proboscis monkeys, etc.*

Cercopithecinae. *Guenons, baboons, mandrills, macaques, etc.*

Generally speaking the colobines are arboreal, leaf-eating monkeys while many cercopithecines are ground-living and omnivorous. Associated with these behavioural characteristics are certain anatomical differences: colobines, for example, have sacculated stomachs quite unlike the simple organs of cercopithecines. In their turn, cercopithecines have their own special dietary specialisation called cheek pouches. These devices are extensible pockets on either side of the inside of the mouth which can be stuffed with food – like shopping baskets – for later consumption (Figure 10).

OLD WORLD MONKEY HOUSE

Cage 15 *Spectacled langur* (*Presbytis ob-scura*). Malay Peninsula. The spectacled langurs (their name derives from the white rings around the eyes) are very typical colobines. Their lips too are white and stand out sharply against the blue-grey colour of the face. Tails are longer than the head and body length combined and the hindlimbs are considerably longer than the forelimbs. Their movement in the trees is tremendously athletic, they bound, leap and drop, and occasionally even swing by their arms through the forest.

Colobines also differ from cercopithecines in the structure of their societies and in the patterns of infant care. The infants of spectacled langurs are bright orange in colour for the first nine weeks of their lives. The contrast with the adult fur colour is dramatic. What possible function can this colour contrast serve?

As far as our understanding goes the explanation is tied up with a very remarkable behaviour pattern of langurs of the genus *Presbytis*. Soon after birth the infant is handed round among the females of its troop who inspect it, pet it and cradle it. Some of these 'aunts' as they are called don't even seem to know which way up the baby should be held. Obviously this social custom is a considerable advantage for survival of the species as it provides a constant training school in baby-care. The contrasting colour of the infant spectacled langur (and other langur species) means the mother can keep her eye on her offspring as it is circulating amongst the 'aunts'. When danger threatens no time is wasted in locating the infant which she will immediately snatch up.

Cage 16 *Guereza. Colobus monkey* (*Colo-bus guereza*). Widely distributed in East Africa from Ethiopia to Tanzania. The colo-bus is an African colobine. Basically it has the typical locomotor and dietetic adaptations of langurs but has in addition a number of peculiar characteristics of its own. The coat is essentially black and white; it is made up of long hairs that form flowing mantles over the back and the rump. The tail is black in its upper regions but thereafter expands into a white, club-shaped extremity. The proportion of white to black on the tail varies between the different races as does the extent of the white mantle (see variation in primates, chapter 2). The infant colobus is wholly white.

The second peculiarity is that – like the South American spider monkey – the colobus has no thumbs. Its very name derives from this special character. The Greek word *kolobos* means maimed or mutilated.

Cage 17 *Diana monkey* (*Cercopithecus diana*). West Africa. The common name for the group, of which the Diana monkey is merely one of the 21 species, is the guenon. Although the Diana monkey has a restricted range, species of guenons are found all over Africa from south of the Sahara to the Cape of Good Hope. One species of *Cercopithecus*, *C. aethiops*, the savanna monkey, is partly ground-living but the rest are more or less arboreal. Diana monkeys are among the most arboreal of all. They are very elegant animals, predominantly blue-black and white in colour with orange saddle and inner thighs; there is a white stripe down the outside of the thigh. The face gives the impression of an inverted triangle, the apex of which is formed by a neat white beard. Diana monkeys are omnivorous. In terms of locomotor pattern they are classed as quadrupeds. Socially they form small groups dominated by males whose status is directly related to their age and experience. Like all Old World monkeys a single infant is the common pattern; twins are a rarity. Gesta-tion lasts six and a half months.

Guenons are the commonest forest monkeys of Africa and although all species have the same general structure they can be distinguished by superficial adornments, usually highly coloured, on the face, thighs and genitals. One is reminded of distinguish-ing regimental flashes of an army corps.

Cage 18 *Military monkey. Patas* (*Eryth-rocebus patas*). Sub-Saharan Africa from

Senegal to the Sudan. The patas monkey (sometimes called the red hussar or the military monkey, on account of the fierce white Colonel Blimp moustache) is a particularly interesting example of evolutionary specialisation. Basically, patas monkeys are guenons, but they are guenons which have long since deserted the forest habitat and taken to a life on the ground. Physically and behaviourally they have undergone remarkable changes which are of immense interest to anyone studying the evolution of primates.

Physically, patas monkeys have evolved in the direction of cheetahs with which they share a common habitat. They have become long-legged and are extremely fast over short distances. Behaviourally, they have adopted some of the patterns of animals living in arid, harsh, predator-ridden domains, forming one-male groups in which the male acts as leader and watch-dog. They have evolved unique ways of avoiding danger. If their speed over the ground is ineffective in shaking off pursuit, then they lie low in the grass to avoid detection; no other primate has evolved this particular predator avoidance procedure. They are also very silent animals unlike their noisy relatives in the forest.

Patas monkeys are large animals with slender bodies, long limbs and moderately long tails. Their dominant colour is rust-red. Males and females have a flowing white moustache which is black in adolescents. The hands are exceptional amongst Old World monkeys in having extremely short fingers, particularly the index finger and thumb, which can be regarded as adaptations for ground-living.

Cage 19 *Pig-tailed macaque (Macaca nemestrina)*. South-east Asia from Burma through Malaya to Sumatra and Borneo. Elsewhere I have expressed my admiration and respect for this resilient primate group, the macaques, by saying that if man is the President of the animal world, surely then the macaque is the Vice-President. Being tough and highly adaptable, macaques have penetrated into all the life zones that primitive man was able to reach. I am

convinced that had not man arrived on the primate evolutionary scene, these ubiquitous creatures would have ruled the world. The only reason that macaques are now restricted to tropical and subtropical environments is that man, fearing formidable competition for food and territory, has driven them out of his principal population centres. Had monkeys enjoyed the same religious tolerance in the United Kingdom (and macaques are known to have reached Britain in the Pleistocene) as they enjoy today in India, then, no doubt, we should be plagued with macaques roaming the food halls of Harrods and Fortnum and Mason – hungry, sacred and unassailable.

The pig-tailed macaques are robust animals with limbs of almost equal length and a short tail carried arched over the back. Their general colour is buff to dark brown; the face is pale and the eyelids conspicuously light. Males are very much larger than females and are further distinguished from them by long, stout upper canine teeth. Females show a conspicuous swelling and reddening of the skin around the genitalia and the root of the tail during several days in the middle of the menstrual cycle when they are in their most fertile phase.

Most macaques move and feed both on the ground and in the trees; the degree to which they are arboreal or terrestrial varies from species to species; pigtailed macaques are among the less arboreal of the genus *Macaca*. Like all members of the subfamily Cercopithecinae, macaques are omnivorous. Social behaviour is based on the multi-male principle in which troop movements and internal discipline are controlled by an élite composed of adult males.

Cage 20 *Olive baboon (Papio anubis)*. West, Central and East Africa. Baboons are even more ground-adapted than macaques although they sleep at night either in trees or on rocky ledges where they are safe from predators; they also feed in trees when certain fruits are ripe. During most of the day baboons range the savannas feeding on grass, roots, insects and even occasionally the flesh of animals such as newborn gazelles.

The olive baboon is a heavily built animal with a compact body and a short tail held high in a croquet-hoop curve. The hands are particularly well developed having a thumb which is second only to man's in the effectiveness of its opposability (see chapter 5). The male baboon is almost twice the size of the female and has strong, aggressive-looking canine teeth which are, however, seldom used offensively except against the cheetah and the lion – the baboon's chief predators.

Olive baboon societies are based on the multi-male principle where discipline, and thus troop stability and survival, are ensured by a hierarchy of dominant males. Position in the hierarchy does not depend wholly on strength and toughness; often experience weighs more. The main role of the dominant males is to maintain peaceful relations within the troop, to protect the young, and to defend the females and infants against attacks by predators.

Cage 21 *Mandrill* (*Mandrillus sphinx*). Tropical forests of West Central Africa. Mandrills are short-tailed forest-living baboons but they possess a number of physical and behavioural characteristics that are specially adaptive for a life spent on the forest floor. For example, one of the problems of forest-living is visibility; in high forest with a closed canopy the intensity of the light on the forest floor is one-tenth normal. Many forest creatures have met this situation by evolving very bright colours that make them easily identifiable. The mandrills are no exception and have emancipated themselves from the overall drabness of their savanna cousins, the baboons, by developing the most striking facial and genital adornments.

The nose of the male mandrill is a brilliant, lacquered, chinese red on either side of which there are extraordinary grooved swellings coloured metallic blue; the cheek tufts are white and the beard yellow or orange. The genital region of the mandrill mimics the facial colours almost exactly: the region around the anus is red, the pubic region scarlet and the penis bright pink; outside this there are blue patches fading to lilac. Thus the male mandrill is equally easily identifiable both fore and aft.

Little is known about the social behaviour of the mandrill, which is extremely hard to study under tropical forest conditions but it seems that they live in groups of twenty or so; in conditions in which hearing is at a higher premium than vision, understandably they rely very heavily on vocalisation to maintain contact and social control.

Monkeys (including prosimians) with tails have been surveyed in a rather limited and superficial way because – fascinating though they are in their own right – monkeys are only bit players in the scenario of this book; the real stars – the superprimates – are the apes and man which we shall be looking at in more detail in the remaining chapters.

Like any human family there are plenty of doubtful characters in the past history of the primate order, creatures which failed to make the grade and now are simply skeletons in the primate cupboard; we shall be considering some of these fossil has-beens in chapters 3 and 4. Some of our more respectable ancestors are also skeletons, but they are displayed in museum show-cases with all the panoply that befits their status as progenitors of the living primates. Among them are certain creatures representative of key groups that constituted the zoological stepping-stones by which monkeys without tails made their way across the turbulent waters of time.

PRINCIPLES OF PRIMATE EVOLUTION

One of the most remarkable features of the primate order is that it displays what one might call 'the staircase phenomenon'. This is illustrated in Figure 6. Each step represents a recognised grade of primate evolution. On the first step are the lemurs, next the tarsiers and so on, each step representing a higher grade of biological organisation than the previous one. On the top step is man. A succession of grades where archaic forms are replaced by new and improved versions is what evolution is all about. There are plenty of examples of this process but perhaps the

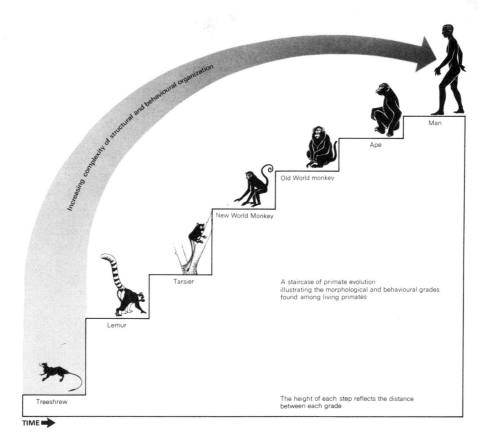

Increasing complexity of structural and behavioural organization

Man

Ape

Old World monkey

New World Monkey

Tarsier

Lemur

Treeshrew

TIME ➡

A staircase of primate evolution
illustrating the morphological and behavioural grades
found among living primates

The height of each step reflects the distance
between each grade

6
The staircase
phenomenon.

most familiar is the evolution of the horse.

Horses developed as a separate line of the ungulates at a very early stage. The tiny 'dawn horse' *Eohippus* (now called *Hyracotherium*) of 50 million years ago is the earliest of the family. *Eohippus* was replaced by the slightly bigger *Mesohippus* which in turn disappeared as a new grade called *Merychippus* became successful. *Merychippus* gave way to *Pliohippus*, and *Pliohippus* to the modern grade of horse *Equus*. As each new horse evolved, the old horses were put out of business – not so with the primates. New grades of primate evolved, but the old grades continued to flourish. Unlike the horses of which there is but a single living genus today, there are approximately 50 genera of primates. (See Figure 7.)

It would be reasonable to ask why the primates are so favoured. This, as the saying goes, is the $64,000 question and it would require the rest of this book to answer it properly. I must ask you to accept a somewhat superficial explanation in view of the fact that we have other, equally interesting, matters to discuss.

Primates are highly adaptable animals, not limited – either by their structure or behaviour – to a single habitat or a single way of life. This means that, as a group, they can make themselves equally at home in the trees, on the ground, on near-vertical cliffs, or scrambling over the roofs of Indian villages. Primates can run, climb, jump, swing, walk on two legs or four and swim. In this versatility lies the secret of their zoological strength. It explains why the early forms of the Order were able to exploit a wide variety of novel environmental opportunities and have thus lived (through their descendants) to tell the tale, while the less versatile horses were replaced when a better adapted horse-type appeared on the scene.

But so far we have merely begged the question. Why *are* primates so versatile? The answer lies in the stock from which primates sprang. The earliest pre-primate mammals were generalised creatures, that is to say they were constructed on a basically simple plan: bodies, tails, four legs of approximately equal length, five digits on each hand and foot, a face with a snout, two eyes, two ears, two jaws with forty-four teeth between them, and a small brain

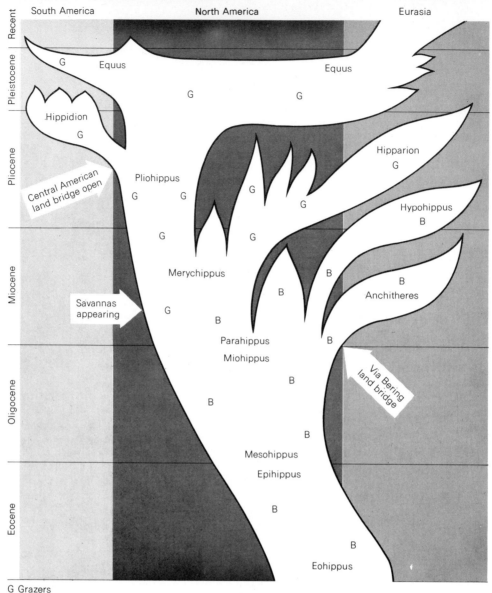

South America   North America   Eurasia

Recent
Pleistocene
Pliocene
Miocene
Oligocene
Eocene

G Equus   Equus
Hippidion
G
Central American land bridge open
Pliohippus
G   G   G   G
Savannas appearing
G
Merychippus
B
B
Parahippus
Miohippus
B
B
Mesohippus
Epihippus
B
Eohippus

Hipparion
G
Hypohippus
B
B
Anchitheres
B
Via Bering land bridge

G Grazers
B Browsers

7 Evolution of horses. Horses started off as browsers but became grazers as grasslands spread during the Miocene. This new habit was associated with structural changes in the teeth which increased in height and in the toes which became reduced in number leading to the single-toed horse of today. (After G. G. Simpson.)

almost wholly given over to receiving and interpreting information derived from the sense of smell. Behaviourally these early pre-primates were quadrupeds, insect-eating and possibly nocturnal; we infer that their social organisation was of the most primitive kind.

From this primitive, uncommitted stock many mammalian lines evolved and eventually developed into the diversity of animal types that we know today. A few examples picked at random include elephants, ant-eaters, hippopotamus, giraffes, bats, whales, dolphins, lions and tigers, hedgehogs, moles, raccoons, bears and giant pandas. Each and every one is different because each and every one is adapted to take advantage of a distinctive aspect of the environment. These aspects are called ecological niches. It is difficult to see how any single mammal mentioned above could have survived if displaced from its particular ecological niche. Niches do not necessarily have to be places or geographical regions. For examples, bats have a wide vegetational and climatic distribution but they are all nocturnal. Giant pandas and anteaters occupy a niche which is more a matter of diet than geography, and lions and tigers can only survive where there is an adequate supply of game. Most of these animals are specialists. Remove the critical element of their lives, be it food supply, an optimum temperature or a particular type of vegetational cover, and they are in danger of becoming extinct. But this is not true of primates; there are very few real specialists in their ranks. Primates are generalists. Geographically, primates can survive in temperate climates or tropical regions; they can eat insects, flesh, fruits, flowers, leaves, roots and tubers, grass and even bark; they can operate at night or during the day. In the face of adversity they will manage to survive somehow.

As examples of the adaptability of primates we can look at the case of the spider monkeys in captivity. These remarkable South American monkeys seldom put foot to the ground in their native tropical forests yet thrive in many zoos in temperate regions in outdoor treeless enclosures where almost all their time is spent on the ground. Another example is provided by the Japanese macaques, short-tailed primates closely related to the rhesus macaques of India. At the extreme northern tip of Honshu, the largest of the four islands that make up Japan, are groups which have been called snow-monkeys. The winter in northern Honshu is severe and winter snows may last for four to five months. Food becomes extremely scarce, and for several months Japanese monkeys may have to ride out the worst of the winter subsisting on the cambium layer of the bark of trees – and little else. This is hardly the image of monkey life that most people have in mind! Adaptability is expressed by the snow-monkeys in several other ways, some physical and some behavioural. An example of a behavioural adaptation is the adoption by one troop of hot-spring bathing in the winter. Japanese monkeys seem peculiarly adroit at inventing new fashions of behaviour. (See chapter 2.) With this background in mind can one be surprised at man's extra-

ordinary resilience and versatility in the most arduous of circumstances?

Primates owe their plasticity in terms of structure and behaviour to two factors: a remarkably generalised body structure and a remarkably specialised brain.

The effectiveness of any brain is dependent entirely on the range and quality of information fed into it. In this sense it is like a computer which has to be programmed. The input comes from the sensory scanning devices of the body, the eyes, the ears, the nose, the tongue and the skin. Together these gather information about the environment through nerve endings which serve the functions of vision, hearing, smell, taste, touch, pain and temperature.

The flashing shuttles of the brain weave the information from the sense organs into a complex tapestry depicting the environmental situation of the moment. Let us take a dramatic example. You are walking down the street and turn a corner and suddenly a series of impulses activated through the five principal senses hits your brain like the staccato stutter of machine-gun fire: the glare of flames, the feeling of heat, the smell of burning, the cries of the trapped and the taste of smoke. Incompletely informative on their own, collectively they are interpreted in the light of past experience as the overall pattern of a house on fire.

An animal deprived of sensory input is a mere vegetable living in a vacuum. The brain depends for its normal operating efficiency on a constant flow of sensory information. As Nigel Calder put it in *The Mind of Man*: 'the brain craves for information as the body craves for food'. Experiments have been carried out where human beings have voluntarily subjected themselves to total sensory deprivation for short periods with unpleasant psychological results including mental disorientation and hallucinations. Such experiments have never been continued long enough for more serious mental disturbances to show themselves. One can guess however at what might happen by studying the consequences of long-term but incomplete sensory deprivation, in political prisoners for example. The technique of

brainwashing starts classically with periods of sensory deprivation alternating with periods of intensive indoctrination and other forms of psychological stress. At a less dramatic level, a state of sensory poverty if not actual deprivation is the lot of elderly people living alone; for them radio and television are not expensive luxuries but mind-saving necessities.

In general, mammals utilise all the five senses but in differing degrees. In most non-primate mammals, smell and hearing are the dominant senses while vision and touch are less well developed. The primate specialisation has been to spread the load more evenly but at the same time to favour vision and touch which are the most precise of all the senses. Furthermore they are complementary senses inasmuch as each only provides half the information required. It is not enough to see an object, it must be handled before its three-dimensional nature can be fully appreciated. The sense of touch, which includes many judgements such as weight, texture, temperature and so on, fills out the visual picture. Dr L. Harrison Matthews in his book *The Life of Mammals* quotes an example of this visual-tactile teamwork. When a friend produces an unusual curio from his pocket you exclaim 'How interesting, let me see it!' but you really mean 'let me feel it' because without the added information that the sense of touch can bring you can contribute little to elucidating its nature. Many people feel a sense of irritation and frustration at seeing a 'two-dimensional' picture, like Van Gogh's *Sunflowers* for example; they feel cheated by the artist because his picture, lacking perspective and depth, is unreal – sunflowers are just not like that in the ordinary world of seeing and touching.

One important aspect of primate evolution therefore has been to adjust the sensory balance by a sort of palace revolution in which visual and tactile elements are ennobled and the sense of smell and hearing dethroned. Expressed in anatomical terms, the physical changes associated with the sensory revolution can be summarised as follows:

Smell. Reduction of the sense of smell is associated with a reduction in the length of the muzzle and a change in the physiological nature of the 'nose' itself, which is 'wet' in mammals and lower primates (prosimians), and 'dry' in higher primates. In the brain the olfactory bulbs, the receptor centres for smell, are much reduced in size.

Vision. Elaboration of the visual mechanism in two different ways: 1. Rotation of the orbits from the side to the front of the face permitting binocular or stereoscopic vision. 2. The development of a special type of receptor cell in the retina which provides the basis for colour vision, and for increased visual sharpness and definition. In the brain such improvements are reflected in the elaboration of the visual cortex which is housed in the occipital lobe at the back end of the brain.

Hearing. The most characteristic external change associated with a reduction of the hearing proclivities of higher primates is the replacement of a thin membranous highly mobile external ear that scans the environment like a radar saucer, with a cartilaginous and immobile appendage that appears to have little function other than to act as a device to keep one's spectacles in place.

Touch. Development of highly sensitive pads on the ends of the fingers and toes associated with a replacement of typical mammalian claws with flat, primate nails. The finger-tips are the most sensitive areas but the rest of the hand also plays an important part in receiving tactile (and thermal and painful) stimuli. A trend is apparent in primates for the pads, typical of the foot of a cat or dog, to flatten out and thus provide a greater surface area for sensory reception. The disengagement of the forelimbs from locomotion as in man, and to some extent in apes, has resulted in the skin of the palms becoming more sensitive. Tactile appreciation is also amplified by an increased mobility of the hand which can wrap itself around an object thus ensuring that a greater surface area is brought into play.

There are major changes in the brain

associated with an improvement of the tactile sense. The area which receives incoming sensory impulses is the pre-motor gyrus, a strip of white matter that adds substantially to the bulk of the parietal lobes of the brain of higher primates.

It must be obvious that a sensory revolution has resulted in a number of important accompanying changes in the anatomy of primates. Among these are: a reduction in length of the muzzle, the configuration of the nose and upper lip, the shape of the ears, the frontality of the eyes, the mobility of the hands, the replacement of claws with nails, the loss of discrete, bulbous, palmar and plantar pads and, finally, a reduction in the hairiness of the body leading, overall, to an increase in the sensory contact between the body and the environment.

The degree to which these characters and their correlates – brain size, posture and locomotor behaviour – have evolved in primates, accounts in no small measure for the physical differences between prosimians and New World monkeys and between New and Old World monkeys that we have observed as we were walking round the primate zoo.

10 Four sharp reasons why monkeys are unsuitable as pets after they have reached maturity. Note the opening in the side of the mouth leading to the left cheek pouch.

11 Some of the 300 pairs of identical twins who
took part in a David Frost TV programme in 1968.

2

Bless 'em all, the long and the short and the tall

'What I want to know,' said the little grey man to the keeper in charge, 'is why your penguins all come in different sizes.' This was a good question if the little grey man had but known it, and it is one that zoologists, geneticists and anthropologists regard as the fundamental question of evolutionary biology. I hope this explains why the first theme to be tackled in this book is variability in animal and human populations.

There is an old north country saying that there is 'nowt so queer as folk' which means, simply, that one should not be surprised at anything that humans do. This perceptive, folksy aphorism is strictly behavioural in its implications; but the Yorkshireman, attuned as he is to the oddities of human behaviour, would not be surprised to hear that folk are equally 'queer' in terms of physical structure. In the human species, variability takes two forms: 1. behavioural and cultural and 2. anatomical and physiological. In this chapter we are concerned with the latter.

The initial premise is that every individual differs from any other individual regardless of consanguinity, sex, race or age. The idea that all babies look alike (or like Sir Winston Churchill, which comes to the same thing) or that you can't tell one Chinaman from another, is simply an ignorant convention, a poor excuse for lack of observation. Paradoxically, even 'identical' twins are different. Mathematically the probability against any two people being exactly alike in every respect can, roughly, be calculated as $2^{12,000}$ (12,000 being a rough estimate of the number of genes in human sex cells). When one realises that 2^{20} represents a product of 1,048,576, one begins to get a glimmering of the magnitude of $2^{12,000}$. The chances of two individuals turning out to be physically identical is not impossible but it is theoretically so remote that it has probably never happened nor is it likely to. When one adds the even more improbable occurrence of behavioural uniformity, one can recognise the absurdity of the basic premise of such romances as Anthony Hope's novel *The Prisoner of Zenda*. On the other hand the superficial identity of the two sets of male twins in Shakespeare's *Comedy of Errors* is acceptable, but not the physical identity of Viola and Sebastian in *Twelfth Night*; twins of different sex can never be 'identical' and may be no more alike than any pair of brothers and sisters.

Twins may develop either from a single fertilised ovum which then divides to form two separate individuals or from two separate eggs. In the first case they are called 'identical' and in the second they are 'non-identical'. With identical twins a single male sperm does the fertilising, and as it is the male component that governs the sex of the offspring, 'identical' twins must necessarily be of the same sex. 'Non-identical' twins may both be males, both females or one of each.

Concordant characters are those that are shared between twins. Concordancy is high in identical twins but is not complete. In some characters like fingerprint patterns and blood groups, concordancy is about ninety per cent. In other traits like the susceptibility to disease it is about seventy per cent. However this figure is a good bit higher than in 'non-identical' twins. It is quite difficult to prove that twins are identical by scientific tests, but there is one rather drastic method that provides proof beyond reasonable doubt. Skin grafts can only be transferred from one individual to another successfully if the genetic patterns of donor and recipient are exactly alike.

At this point we must make a distinction between two types of characters: those that an embryo acquires as a result of the blending of the maternal and paternal genes, and those that it acquires throughout life as a result of environmental influences. These two sets of characters are sometimes referred to as genetic and acquired and sometimes – when behavioural characters are at issue – as inherited and learned. The two forces at work producing these characters have somewhat picturesquely been labelled 'nature' and 'nurture'. The much overrated and overstated 'nature-nurture controversy' concerns the relative importance of these two processes in shaping an adult human being. Nowadays the consensus opinion is that both processes play their part – an individual is neither wholly a product of his genes nor of his environment. The 'controversy' has little meaning for biologists; for sociologists, who now seem to be discovering it for the first time, it is fast becoming a burning issue particularly in the controversial areas of race and education.

Identical twins start off on an equal footing as far as their genetic characters are concerned but from then on they are subject to the additional influences of the environment which starts to have its effect in the womb. Even at birth, differences are already established as the result of the vicissitudes of nine months of intra-uterine development. External influences such as the adequacy of the blood-supply, the position of the foetus in the womb and the mechanics of birth, ensure that the two individuals are no longer 'identical'. However, they are still identical enough for twin-studies to be a most important technique for studying the effects of 'nurture' as opposed to 'nature'. Given genetic identity, it follows that all observable differences must be environmental in origin; and if, as sometimes happens, 'identical' twins have been separated from birth on a prince-and-pauper basis, then the observable differences in their intellectual and social attainments are of immense value in the study of environmental influences.

Geneticists employ the terms genotype and phenotype to describe, respectively, the genetic composition and the characteristics of the organism; the phenotype is composite, being made up of genetically and environmentally induced characteristics.

While phenotypic variation in man and animals is something tangible that can easily be demonstrated, genotypic variation is cryptic, hidden away in the cells of the body. The chemical constituent of genes, arranged in a double spiral form (the 'double helix'), is known as deoxyribonucleic acid (DNA) and is the chemical basis of what we refer to as 'life'. DNA synthesises amino-acids which in turn combine to form the proteins from which the body is built up. The miraculous molecules of DNA are made of chemical sub-units, and heredity is determined by the order in which these sub-units are arranged on the gene.

The genetic pattern provides a set of instructions (in the form of a chemical code) which determines the exact form that a new individual will take. The possible combinations are astronomical in their magnitude. Each individual animal is therefore 'custombuilt' but, unlike bespoke suits, the customer is never consulted; as we see it the process is entirely random. The accidental changes – or mutations – in the genotype lead to what we call variations in the phenotype; variations, in their turn, having passed through the sieve of natural selection, provide the raw material for evolution. Mutations are sudden and relatively rare events that are probably induced by external influences such as X-rays, ultraviolet rays, excessively high temperatures, gamma rays, neutron bombardments (man-made or otherwise), and by certain chemical agents. In this sense, mutations are environmental in origin. Most major mutations are lethal and lead to the destruction of the embryo long before it is subjected to the 'sieve' of natural selection. Major mutations probably play little direct part in the normal evolution of a species but may have played a significant role in some of the major evolutionary shifts of the past as, for example, the emergence of land-animals and, much later, the extinction of the giant reptiles.

The principal source of phenotypic variability is in the process of recombination which refers to the mixing in the offspring of the gene contents of the parents' chromosomes when they come together during

Open savanna (East Africa)

Rain forest (Borneo)

Japanese macaques from northern Honshu. *Left:* Hot-spring bathing. *Right:* Riding out the snowy winter in spite of food in short supply.

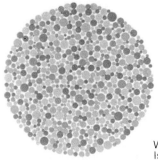

What the male motorist would see with defective perception of red. Above is an example of Ishihara colour charts; the figure 6 is clear to those with normal vision.

sexual reproduction. Although mutations are the ultimate cause of variability, they may lie fallow for many generations before some quirk of recombination brings them to the fore.

THE FLOW OF GENES

In one sense the whole life style of animals is a preparation for ensuring the continuity of their species by the reproduction of their kind. Animals that do not reproduce are biologically irrelevant for their particular genetic pattern is gone for ever. Theoretically the label of 'biological irrelevancy' could be applied to modern man as well as to animals, but in practice the human way of life has set its own values, and successful breeding in terms of numbers of offspring does not figure very high amongst its priorities. The theory of biological irrelevancy would certainly have applied to early man living a tribal existence where populations were small, and contact with other human populations few and far between. It would also have applied in moments of crisis to small groups like Neanderthal Man isolated by the encroaching glaciers of the last Ice-age. It would also have applied to populations decimated by the sweeping epidemics of the black death in medieval times.

More important than the loss of population numbers is the loss of genes. Interbreeding populations of animals possess, collectively, a stockpile of genes and it is upon this source that future populations depend for their further evolution. The bigger the stockpile, the less likely it is that a population will become evolutionarily static. In small populations the same old genes are just going round and round, but in large populations there is an almost infinite potential for change. New 'blood' – or rather new genes – are needed for a population to undergo evolution, and it is for this reason that 'biological irrelevance' is a critical factor in small populations but unimportant in large ones.

The loss of genes is no problem to a species like *Homo sapiens sapiens*, or modern man, which is widely distributed over the habitable parts of the globe. Geographical isolation, so important in animal populations, has relatively little meaning in the modern human context. Nowadays the only physical barrier between the mating of a Briton and a Hottentot is the price of an airline ticket; the barriers limiting the flow of genes between human populations owe little to geography. The principal barriers to random mating are racial, cultural or economic, powerful enough deterrents one might think, but, in the event, not powerful enough to prevent extensive intermingling of the available *H.s.sapiens* genes. Thus mankind can afford to support genetically non-contributory individuals. This is partly what is meant by civilisation. As a consequence, modern human societies find room for a certain proportion of religious celibates, sexual iconoclasts and deviationists without seriously endangering the future of the species.

The genetic composition of an organism (human or otherwise) is a cumulative index of the history of the species. The genes of the earliest human ancestors of two million years ago, for instance, are still to be found in the human stockpile. Some are even recognisable for what they are. Genetic characters associated with Neanderthal Man, who vanished as a distinct population 30,000 years ago, can be still recognised today, although next time you sit next to a hairy, brutish monster of a man in the bus, don't assume that what he is carrying in his case is a club studded with vicious-looking knobs – it's more likely to be his guitar!

The genes of modern man derive from ancestors that are far more ancient than cavemen. We all have apes in our attics, monkeys on the mezzanine and lemurs in the lounge. For brief periods during our development we even possess the tail of a reptile and the gill-slits of our fishy forebears.

NATURAL SELECTION

Genes are the ultimate source of evolutionary change but the new variations they engender through mutation or recombination during the blending of the male and female sex cells have yet to undergo a rigorous process of selection before becoming built into the genetic pattern of the species. This is the process that Charles Darwin called natural selection. When he visited the Galapagos Islands on his voyage on HMS *Beagle* Darwin observed, for instance, that within the fourteen species of finches that inhabit this archipelago there was a great variety of different ecological types, each of which was adapted to a particular feeding niche. He realised that species could show variability within their ranks and that such variability provided the raw material of their future evolution.

Darwin also observed that there was a need for a natural force to control the size of animal populations. Without such a mechanism the number of animals would soon outrun the food resources available; he further observed that many more animals were born than ever reached adulthood. Clearly there had to be some sort of weeding-out process or selection operating in nature that eradicated the unfit while preserving the fit. It was these observations amongst others gleaned from his travels that led Darwin to formulate his theory of natural selection.

Before discussing this fundamental principle, an explanation of the rather curious words 'fit' and 'unfit' might be helpful. 'Fitness' for Darwin was the sole criterion for individual survival; in fact his own phrase 'survival of the fittest' came to be regarded as the epitome of Darwinism, but like so many catchphrases it opened itself to a great deal of misrepresentation. 'Fit' does not mean strong or healthy, it means suitable or well-adapted. In the Darwinian sense the 'fittest' animals were those, irrespective of their size and strength, that left the largest number of surviving offspring. In this sense rabbits are 'fitter' than lions although this analogy has little meaning as rabbits and lions live in totally different environments and are not in competition with each other; however the point is that strength has nothing to do with 'fitness'.

Natural selection is the process by which new characters, the outcome of change or rearrangement of the chemical units of the DNA molecules of the genes, are tested for 'fitness' or adaptive value for the individual. The test bench for all such genetic innova-tions is the environment. Natural selection operates through the environment. To take an extreme example, animals with thick coats and a thick layer of subcutaneous fat can survive (and breed) in arctic conditions; animals without these adaptations are unable to survive and are eliminated without leaving any offspring; in terms of the environment the 'fittest' have been selected.

When a new aeroplane is developed it is subjected to exhaustive tests on the ground and in the air to evaluate its structure, function and behaviour. If it fails these tests it is quietly dropped and no one hears any more about it. If it passes, then it is granted a certificate of airworthiness and takes its place among existing species of aircraft; once established, it 'breeds' other types – Mark II, Mark III and so on. In other words the new aircraft starts its own line of evolution as a consequence of aeronautical selection. A new species of aircraft has been born.

MEANING OF SPECIES

We have talked a lot about species and now the time has come to explain what they are and how they fit into the evolutionary picture. By definition, a species is a collection of individual animals (or plants) that are capable of interbreeding and producing fertile offspring. Theoretically, all members of a single species can mate successfully with one another but not with members of any other species. As an example, while all dogs (Yorkshire terriers and great danes, size notwithstanding) of the species *Canis familiaris* can interbreed, they cannot mate successfully with the wolf (*Canis lupus*) or with the Indian jackal (*Canis aureus*) which belong to different species.

In spite of the rule book, breeding can and does occur between different but closely related species under certain conditions. The progeny of cross-mating are called hybrids but are usually sterile. Lions and tigers may hybridise producing a cross called a liger when the male is a lion, or a tigon when the male is a tiger; so may mares and male asses, the product being a mule; dogs and foxes are also reputed to hybridise. Although it seems very unlikely that any new species of animal have been produced in this way, hybridisation is a common source of new species where plants are concerned.

Hybridisation in animals takes place when the normal geographical and behavioural barriers are broken down. Captive animals which are living abnormal lives and mixing with animals they would not normally meet in nature are prone to cross-breed. Hybridisation also occurs in nature where the geographical ranges of two species of the same genus abut on one another. A good example of this occurs between two species of baboon (*Papio hamadryas* and *Papio anubis*) whose territories meet along a broad front in eastern Ethiopia. Here a hybrid-zone, as it is called, has become established.

A model of the evolution of a species in a simplified style is shown in Figure 12. The diagram shows the fragmentation of an original population into two sub-units that, while remaining apart, occupy broadly the same ecological niche. There is no interbreeding between these two sub-units simply because they are not in contact; the result is that their respective genes do not mix. Minor structural and behavioural differences develop in the two sub-units as a result of their genetic and ecological separation. At this point in the model an 'agent' is introduced which separates the two sub-units and any flow of genes between them that might conceivably have existed before is completely blocked; it is wholly immaterial what exactly the 'agent' is, it might be an uncrossable river, a mountain range, or a desert. Now follows a period of isolation when the genetic differences become more pronounced, sufficient for the two sub-units to be called subspecies.

While the geographical barrier persists and time goes on, the two subspecies will eventually become separate species. Disappearance of the barrier will once again bring the two original sub-units into physical contact but by now they are separate species and their genetic patterns so different that they can no longer interbreed; a reproductive barrier has been set up. A certain amount of hybridisation is still possible but it is of negligible significance for the evolutionary future of the two species.

From what has already been said it will be becoming clear that species can be defined in terms of reproductive isolation. In 1940 Ernst Mayr of Harvard University provided the following definition: *Species are groups of actually or potentially interbreeding natural populations which are reproductively isolated from other such groups.*

Figure 12 also provides a model to show how major environmental differences may lead the two new species along quite separate paths with the result that they evolve into quite distinct types of animals. By now they are so different in their structure and behavioural habits that a classifier seeing them for the first time would group them into two different genera or possibly even into two different families. Linnaeus's chief contribution to natural science was to introduce a system of classification of plants and animals; a system which, with additions and modifications, we still use today. Linnaeus recognised the following categories: *Class, Order, Genus, Species* and *Variety*. The present system is more comprehensive and runs as follows:

Kingdom Animals collectively

Phylum Major subdivision of animals: i.e. those with backbones and those without

Class Fishes, birds, reptiles, mammals, etc. are in separate classes

Order Mammalian orders include the carnivores, the insectivores, the primates and so on

Suborder Among the primates, monkeys, apes and man are in one suborder, lemurs, and bushbabies in another, tarsiers in a third

Superfamily Old World monkeys are in one superfamily and apes and man in another

Family Man and his ancestors form one family; apes and their ancestors form another

Genus Early men and modern men are placed in different genera (i.e. *Australopithecus* and *Homo*)

Species There have been a number of different species of men in the past, but now there is only a single species (*Homo sapiens*)

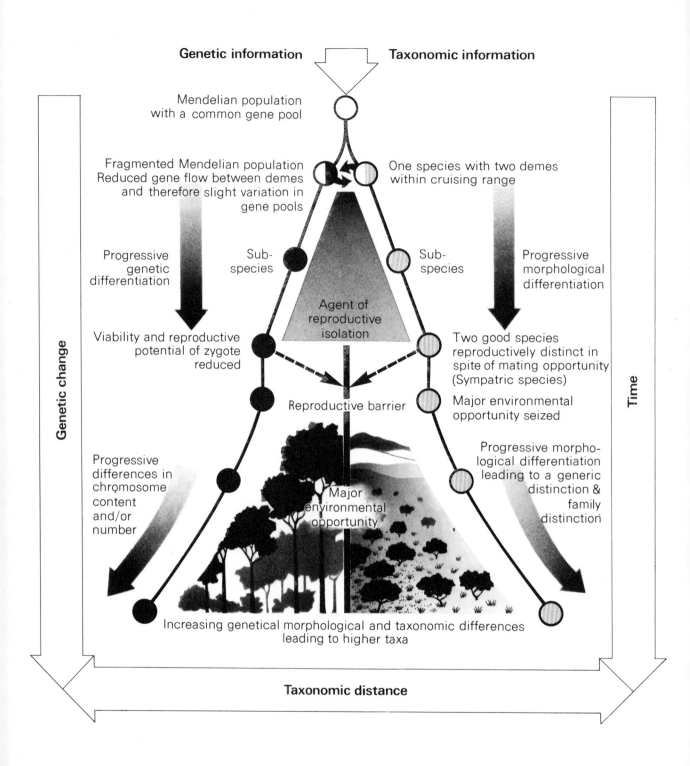

Genetic information **Taxonomic information**

Mendelian population
with a common gene pool

Fragmented Mendelian population
Reduced gene flow between demes
and therefore slight variation in
gene pools

One species with two demes
within cruising range

Progressive
genetic
differentiation

Sub-
species

Sub-
species

Progressive
morphological
differentiation

Agent of
reproductive
isolation

Viability and reproductive
potential of zygote
reduced

Two good species
reproductively distinct in
spite of mating opportunity
(Sympatric species)

Reproductive barrier

Major environmental
opportunity seized

Progressive
differences in
chromosome
content
and/or
number

Major
environmental
opportunity

Progressive morpho-
logical differentiation
leading to a generic
distinction &
family
distinction

Increasing genetical morphological and taxonomic differences
leading to higher taxa

Genetic change

Time

Taxonomic distance

12 A model to show one way in which species and higher categories can evolve from a single population.

Subspecies An important category in animals. In man subspecies are broadly (but not exactly) equivalent to geographical races. Modern man is called *Homo sapiens sapiens*; Neanderthal man, now extinct, was of a different race known as *Homo sapiens neanderthalensis*.

Additional categories such as infraorder, subfamily, subgenus are sometimes used but need not concern us here.

NATURAL SELECTION IN ACTION

Most people who have heard of natural selection imagine it to be a hypothetical process rather than what it is, a real event going on under our very noses. Admittedly it is not very easy to observe natural selection happening, so one tends to think that, whatever it was, it is all over and done with now. Even in man this is far from the truth. As environments change, new modifications of previously successful adaptations need to be up-dated, and as environments are in a constant state of flux, no adaptation can be regarded as a permanent fixture in a species. Sometimes environments change so suddenly and completely that old adaptations become redundant. They either disappear, like the so-called pineal organ (or third eye) of reptiles, become modified, like the gill-slits of fishes, into other structures, or are retained as useless etceteras like the appendix in man. Structures that have no apparent function in a living animal but are believed to have played a role in the economy of its ancestors are called atavisms. Most environmental changes however are gradual and thus most adaptations become imperceptibly modified.

One of the best documented examples showing that natural selection is still an ongoing affair in man is the weight of the human infant at birth. There is an ideal weight for new-born babies which gives them the best chance of survival. The actual weight varies in different parts of the world. There is also an average figure based on the weight of samples of babies born. Natural selection operates to ensure that the average weight corresponds closely to the ideal weight; in this way the human new-born is given the best chance for survival. The way natural selection works in this context is that grossly overweight or underweight babies are eliminated in one way or another during infancy and early childhood, usually because of their greater susceptibility to infectious diseases. In this way the genetic pattern of the parents that has resulted in an abnormal birth-weight in their offspring is not perpetuated. The particular genes responsible for overweight or underweight offspring are gradually reduced in frequency and eventually eliminated from the human genetic stockpile. Of course, the death of over- or underweight infants is not inevitable by any means. Medical care does much to prevent it happening, but it does mean that over- or underweight infants are at a very slightly greater risk than infants of the ideal weight. A character only has to possess an infinitesimal advantage or disadvantage for natural selection to take effect.

We read often enough these days of multiple births following the use of fertility drugs. The classic picture is four or five infants born prematurely; one dies, then another dies and often a third and fourth finally succumb. One can hardly envisage a better practical demonstration of Darwin's axiom of survival of the fittest than this pattern of the elimination of these artificially induced misfits.

Variation in average birth-weight all over the world is well known. In Britain and most European countries it is 7.50 lb, in South-East Asia 6.87 lb, and in India 6.48 lb. That this seems to be largely a genetic effect and not simply due to better food and improved antenatal care and so on, is demonstrated by the fact that immigrants to the UK from India, for example, bring their average birth-weight tendencies with them. Records however are insufficient at present to determine whether or not a gradual shift towards the average birth-weight characteristic of the host country will gradually supervene amongst immigrants. If there is an environmental factor operating in addition to the genetic one, such a shift is to be expected.

Heights of immigrants are better documented. Two American anthropologists, Shapiro and Hulse, have shown that among Japanese immigrants to Hawaii the first generation sons were 2.5 per cent taller than their fathers; they have also increased considerably in weight. Clearly environmental factors such as better food and less debilitation from diseases and parasitic infections were the principal factors in this example.

Another example of natural selection in action is the well-known case of the 'sickle-cell gene'. At least twenty per cent of modern West Africans carry this gene, but in only about nine per cent of the descendants of West African slaves imported into the southern states of North America can it be identified. The gene, which causes the production of an abnormal haemoglobin in the blood, is debilitating and may be lethal, but it bestows a remarkable and – until recently – a quite unsuspected advantage. It protects against malaria and so is important in West Africa. Malaria is not endemic in the United States, thus the possession of the gene has no redeeming features for residents of that country. Sufficient sickle-cell carrying blacks die before reaching child-bearing age for the gene to be gradually eliminated from the population by natural selection. Already in the course of about 200 years its frequency in the United States has been reduced by eleven per cent.

All the examples quoted above, from which we infer that natural selection is a here-and-now phenomenon, depend on statistics for their demonstration. It would clearly be much more convincing if we could see natural selection in action with our own eyes. It so happens that we can.

Up to 1848 the peppered moth (*Biston betularia*) was well adapted for protecting itself from predation by birds by resting on the trunks of trees where its greyish-white colouring blended perfectly with the light coloured bark and lichens. By 1850 a new variant of peppered moth had appeared; the new variety is darkly blotched all over wings and body with a black pigment called melanin. By 1900 the melanistic form of peppered moth had replaced the lighter

variety in the Black Country areas of the Midlands in the ratio of 99:1.

The environmental change that had brought this about was the Industrial Revolution. The inventions of Richard Arkwright, the Bolton barber who introduced the water-propelled cotton-spinning machine, and Samuel Crompton, who invented the spinning jenny in 1779, not only altered the social environment of eighteenth- and nineteenth-century Britain but, as we now appreciate, played havoc with the physical environment. The smoke and soot of 'the dark satanic mills' settled on the green, unpolluted fields and woodlands of England and turned them black.

The principle underlying the success of the melanistic moths is simply one of camouflage. A white moth on a white ground is nearly invisible, but it stands out starkly when placed on a dark background; similarly a black moth on a dead black surface fades into the background, but cries out for attention on a white surface. Translating these simple facts into the natural situation, the melanistic moth has a much greater chance of survival against the background of black, sooty bark than has the light-coloured species which is subject to predation by birds. Thus in the polluted woodlands of the north Midlands the melanistic variety of moth tended to survive and breed in much greater numbers than the light form. The agents of natural selection in this case were the architects of the Industrial Revolution, and the hatchetmen were the birds.

Dr H. B. D. Kettlewell and Professor Niko Tinbergen, zoologists from the University of Oxford, recorded on film the actual process of selection. The film shows that when both varieties are placed on the dark-coloured trunks of trees, the birds (redstarts, spotted flycatchers and nuthatches) swoop down and pick off the light-coloured moths ignoring the melanistic variety which survived these predatory attacks.

A heartening postscript to this fascinating story with its depressing overtones of a polluted countryside is that in those very same areas the light-coloured moths are

13
The melanic variety of the peppered moth and the original form.

beginning to come back again as anti-pollution measures begin to take effect.

Behavioural Selection: So far we have been discussing selection in action in terms of structure, but the principle is equally applicable to behaviour, since natural selection is wholly concerned with the interaction of the living animal and its environment. In the aeroplane analogy quoted above, aeronautical 'fitness' can only be determined when the aircraft is airborne – when it is operating in its natural environment. But of course without structure there can be no behaviour so one is simply dealing with two different manifestations of a single problem.

It is always more telling to make one's point with a familiar and domestic example than with an esoteric one, and what could be more non-esoteric than milk bottle tops. You may well ask what milk bottle tops have

got to do with natural selection but if you reflect for a moment on the occasions when on taking in the milk in the morning you have been infuriated to find a jagged puncture hole in the top and half an inch of cream missing, you will begin to get an inkling of what I am talking about. I am talking about tits, particularly the blue tit and the great tit – charming birds but shocking scavengers.

This behavioural pattern of tits was first recorded in Southampton in England in 1921, at a time when milk bottle tops were made of cardboard with little tabs to lift them out by. The fairly recent introduction of foil tops no doubt gave these avian bandits an added impetus to pursue their life of crime by making their task even easier; one sharp peck on the drum-tight top and they were through! Infuriating as this habit is to housewives, it provides confirmatory evidence to behavioural scientists of the rapidity of the learning process in animals and the speed with which a cultural innovation can spread through a population. Clearly milk bottle tops have no place in the genetic code of the tit family but the behavioural pattern of pecking for a food reward certainly does; a precisely similar technique is used for getting at insects under a layer of bark. The innate or inherited component of their behaviour has simply been redeployed from bark to milk bottle tops. Tits display a degree of behavioural 'plasticity' that we mistakenly regard as the prerogative of the higher mammals such as primates.

Let us take another example of natural selection in action in a behavioural context, and this time in one of these higher mammals – the macaque monkey. The subject is potato-washing; the potatoes are sweet potatoes and the locale is a sandy beach on an off-shore Japanese island.

The island in question is called Koshima,

14
Behavioural evolution in action.

15
Sweet potato-washing. Japanese macaques.

16
A Japanese
macaque,
its hands
occupied
with carrying
sweet
potatoes,
driven to
walking on
two legs.

the macaque monkey is a Japanese macaque (*Macaca fuscata*) and the sweet potatoes were used to lure the monkeys from their normal habitat, of temperate woodland forest, down to the beach where they could be scientifically observed. It all started with a monkey called Imo, a young female about one and a half years old. She was the first of her contemporaries and elders to get it into her head that if she carried the sweet potatoes into the sea and washed them, their tastiness could be improved. I have already transgressed far beyond the legitimate limits of inference by imputing motives to Imo that cannot possibly be proved, so I might as well go the whole hog and suggest that her reason for so doing was to remove the sand or, alternatively, to add the piquancy of a touch of salt.

Motivations apart, the measure of Imo's cultural breakthrough was that after nine years seventy-three per cent of the Koshima macaques had adopted the habit of potato-washing. The sequence of cultural transmission was as follows: Imo passed the habit to her mother who passed it on to her 'best friend', another adult female who, in turn, was imitated by her daughter. Imo at the same time was influencing her juvenile contemporaries just as, in parallel, Imo's mother's best friend's daughter was doing; in this way the potato-washing habit spread rapidly throughout the troop. Imo, clearly a most enterprising young female – a kind of simian Mary Quant – then introduced a new craze into the Koshima troop called 'wheat-washing'. The Japanese experimenters were in the habit of throwing handfuls of wheat grains into the sand to keep the macaques busy picking up the grains while observations of the composition of the troop and the relationships between its individual members were being made. To begin with this device worked, but then Imo, the incorrigible, started thinking again and one day she scooped up a handful of sand and wheat grains and threw it in the sea; the heavy sand sank and the light grains floated and a new cultural habit was born.

Japanese primatologists, who have been studying these monkeys with intensity and enthusiasm for the last twenty years or so, report many similar incidents. In each case the new habit was introduced by a juvenile and the spread was in two directions: to the mother and through her to her contemporaries, and to the inventor's contemporaries within the same 'play group'. What is particularly interesting about all these incidents is that adult males do not figure in the story at all. They appear to form a conservative cadre sitting on the sidelines bristling with disapproval at the permissiveness of the females and the younger generation.

In the discussion of natural selection of behavioural characters, I have to admit that my two examples provide possible models for this phenomenon rather than proven examples. In fact, we have no evidence that these particular expressions of learned behaviour have, through natural selection, led to an increase in fitness of the species concerned, but possibly this is because insufficient time has elapsed for genetic changes to become established. Perhaps these behavioural characters will, in the long run, leave no record in the genetic code. My justification for quoting them is that they illustrate the way natural selection works in respect of behavioural characters. Supposing that the cream at the top of milk bottles, or sweet potatoes, became the principal food supply of tits and Japanese macaques respectively, then one might expect eventually, for the sake of argument, to observe the evolution of a sharper beak in tits, and an improvement in the manipulative ability of the hand in the Japanese macaque.

We can now see that variations are the raw material of evolution. Whether genetic or acquired, inherited or learned, they are subject to natural selection at the test-bench of 'fitness'; those traits that further enhance the survival of the species (judged in terms of reproductive success) are incorporated into its genetic bank account to be drawn on by future generations. One thing should now be clear: variations that become fixed in the genetic code provide some survival advantage for the possessor but we may not always be able to interpret their significance. With this in mind we should be in a better position to examine some examples of variability, first in mammals in general, and then in a particular mammal, man.

No two humans are exactly alike, nor are any two animals. A population of animals normally shows variability within its ranks. The source of the variation is genetic and lies in the particular gene pattern of any given individual. That being so, its descendants will inherit that particular variation in accordance with the strict laws of heredity. If the original population breaks up into two or more sub-units as illustrated in Figure 12, one can assume that a particular variation will be equally represented in the two sub-units. If, now, a geographical barrier is introduced which isolates these two sub-units from each other so that no further inter-breeding can occur, different types of variation are likely to manifest themselves dominantly in the two sub-units as a result of the due process of natural selection. Supposing the variations concern colour patterns of the hair covering the body, it will follow that the two sub-units will eventually come to look rather different – one will have broad stripes, for example, and the other will have narrow stripes. This difference may be enough for zoologists to recognise two different types of the original population. If the original population was a recognisable species, then the two different types can be called distinct subspecies.

VARIATIONS IN MAMMALS

Apes. Variations within a single inter-breeding population of higher primates are particularly well marked. No one can quite explain why the apes (and indeed man) exhibit this characteristic so strongly. Perhaps it is because these creatures have been so

17 Individual variation in mammals. Variability seems to be more marked in chimpanzees than in white-faced sheep.

Variation in
the skin
patterns of
giraffes:

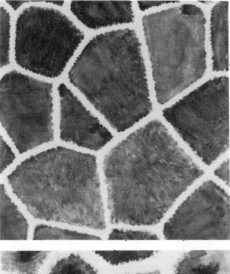

18
Reticulated
pattern.

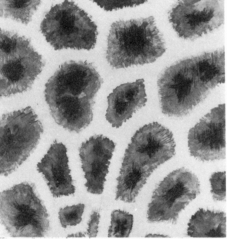

19
The nuclear
pattern.

20
The irregular
leaf-pattern.

closely studied that we are highly geared to spot minute differences; if leopards had been subjected to the same critical analysis perhaps we would change our view of the unique variability of apes. On the other hand, it is possible that the striking variability of apes and man has some significance in its own right; it could be related to the greater demand on the recognition of the individual in the social lives of these creatures.

Variation extends through all systems of the body. Some years ago when I was studying chimpanzee skulls at the Powell-Cotton Museum, Birchington, Kent, I laid out their remarkable collection of 200 skulls in one line extending the length of a long gallery. As I walked slowly up and down I was astonished that while there was overall similarity, each skull was as different as the faces of human beings seen in their serried ranks at football matches; some were long and thin, others broad, and some were squashed with heavy beetling brows. I couldn't find a matching pair among them. The same degree of variability is seen in gorilla and orang-utan skulls.

Jane van Lawick-Goodall, whose studies in the Gombe Stream Reserve in Tanzania have provided so much new knowledge about chimpanzees, makes a great point of this variability in the population she studied. In fact she gave each individual a name according to their facial idiosyncrasies or behavioural oddities.

Hoofed Mammals. Variation between different geographical races of a single species, which normally do not interbreed, is exemplified by the giraffe. The giraffe's variability in coat pattern is a striking example; there are three main colour patterns with an infinite variety of intermediate types. To the north of East Africa in Ethiopia and Southern Sudan the crisp outlines of the reticulated pattern, looking like a jig-saw puzzle ready to be assembled, is by far the commonest. Further south in Uganda and Kenya the crisp patches become blotchy and irregular in outline and develop dark, star-shaped nuclei; still further south in the

Serengeti, the irregular leaf-pattern of the Masai giraffe predominates. Exactly what is the advantage of such variability is not understood; one might suppose that it functions as camouflage and therefore is intimately related to the nature of the vegetational patterns of the regions in which each race lives. In general terms, the patches of a giraffe are seen as devices to break up the animals' outline.

The long neck of the giraffe is generally held to be a feeding adaptation which allows these graceful animals to browse on the foliage of trees which other browsers cannot reach; the long legs of the giraffe complement this survival trick. The okapi, a forest version of the giraffe, is also a browser, but as the foliage grows nearer the ground its adaptations do not need to be so extreme.

Coat variation amongst zebras is of a different order. Unlike giraffes, which constitute a single species with a number of races, zebras come in three separate species; although they are alike in that all have stripes, they differ sufficiently in other ways to merit taxonomic separation. Variation *between* species is only an extension of variation *within* a species, the same genetical mechanisms underly both, thus it is a matter of 'a distinction without a difference'. Your basic zebra is a horse with stripes. In fact it belongs to the same genus, *Equus*. It is fairly small, not exceeding 12 hands, has a large head, a short erect mane, large ears and narrow hooves, and is altogether a very well-adapted creature for a life spent on the grassy savannas. In Figures 21–23 the principal varieties are illustrated: Grevy's zebra, Burchell's zebra and the mountain zebra. The Grevy, the most northerly form in Africa, has the most dramatic stripes of all; they are narrow and black and show up strikingly against a white background; the underbelly is pure white and the ears are shaped like large, rounded radar saucers which collect and concentrate sound. The mountain zebras, from the south and south-west, have prominent stripes on a creamy ground, a thicker coat, a dewlap hanging from the neck and – most striking of all – a grid-iron pattern of stripes on the rump. In

The three species of zebra:

21
Grevy's zebra.

22
Burchell's zebra.

23
Hartman's **or** mountain zebra.

one race of Burchell's zebra, each stripe has a faint 'ghost' stripe of lighter colour. Once again, the nature of the increased 'fitness' that these variations reflect must be taken on trust.

Man

Man comprises a single species, *Homo sapiens*, and for most scientific purposes can be treated as a single interbreeding population, although there are a number of isolated communities still to be found in many parts of the world.

Human diversity expresses itself in two forms: continuous and discontinuous. Typical of a continuous variation is stature which forms an unbroken sequence of heights from the shortest to the tallest. Discontinuous variations, as their name implies, are disjointed; for instance human blood groups of the ABO system can be either A, AB, B or O; there are no intergrades.

The basic variation of course is sex, and if one discounts the various varieties of physical intersex, which are developmental abnormalities, sex is clearly a discontinuous character. Every child that is born is either male or female. W. S. Gilbert, the librettist of Gilbert and Sullivan, expressed this biological fact in terms of political parody:

> *I often think it's comical*
> *How Nature always does contrive*
> *That every boy and every gal*
> *That's born into the world alive,*
> *Is either a little Liberal,*
> *Or else a little Conservative.*

Sex is determined by a pair of chromosomes. Two similar X chromosomes are carried by the female ovum and two different X and Y chromosomes by the male sperm. The ovum may be fertilised by either type of sperm. Thus, two possibilities exist:

sperm X + ovum X = XX. The baby is female

sperm Y + ovum X = XY. The baby is male

It is the male of the mating pair which determines the sex of the offspring. Sex is the variation of variations and perhaps should not be considered in the same context as stature and blood groups. So now let us look at some of the others. Many human characteristics are polygenic, that is to say they are influenced by many genes; all the characters discussed below fall into this category.

Blood groups. Although we refer to ABO blood groups as if they were the only blood system, there are in fact many others including the Rhesus system which is so important in certain diseases of the new-born. However the ABO blood groups were the first to be described and are the most familiar to most people. Geographical distribution of ABO groups is of great interest to anthropologists. For instance in eastern European populations there is a preponderance of Group B while in North American Indians of pure stock Group B is wholly absent; these people show a 99.9 per cent frequency of Group O. Group Bs are unknown, too, among Australian aborigines. As a rough guide to the overall frequency of each group, the percentage breakdown for western European populations is as follows: A: 42, B: 9, AB: 3, O: 46. In terms of blood transfusions Group AB are known as 'universal recipients – they can receive blood from individuals of any other group without a disastrous clotting of the corpuscles taking place. Group O individuals are 'universal donors' and can give blood to all other groups, with certain safeguards.

For anthropologists the frequency of the blood groups provides insights into the past history of populations. Blood groups are purely genetic in origin, there is no question of any environmental effects, therefore they act as genetic markers flagging the flow of genes between one geographical region and another. For example within the United Kingdom there is a higher frequency of Group A in Pembrokeshire than anywhere else in the British Isles. This suggests to demographers that these genes were derived from the Viking invaders who are known to have established settlements in this area. In regions associated with the survival of Celtic stocks in Scotland and Wales the frequency of Group B is statistically higher than in other parts of the country.

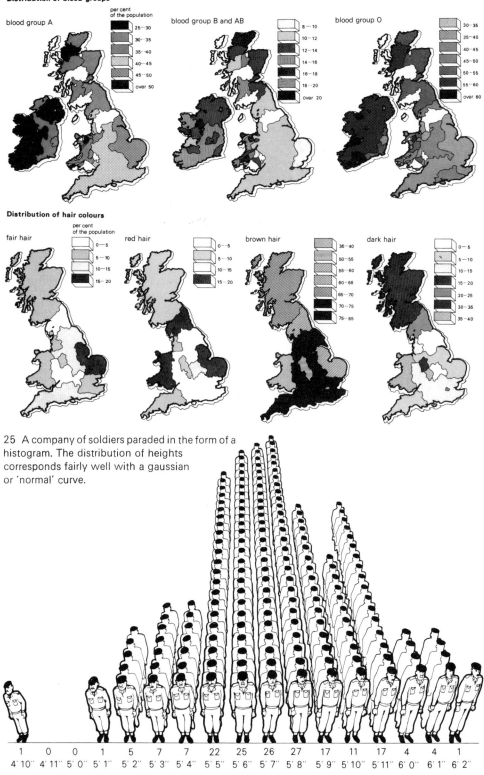

24 Blood groups and hair colours in the British Isles.

Distribution of blood groups

blood group A

per cent
of the population
25—30
30—35
35—40
40—45
45—50
over 50

blood group B and AB

8—10
10—12
12—14
14—16
16—18
18—20
over 20

blood group O

30—35
35—40
40—45
45—50
50—55
55—60
over 60

Distribution of hair colours

fair hair

per cent
of the population
0—5
5—10
10—15
15—20

red hair

0—5
5—10
10—15
15—20

brown hair

35—40
50—55
55—60
60—66
65—70
70—75
75—85

dark hair

0—5
5—10
10—15
15—20
20—25
30—35
35—40

25 A company of soldiers paraded in the form of a histogram. The distribution of heights corresponds fairly well with a gaussian or 'normal' curve.

| 1 | 0 | 0 | 1 | 5 | 7 | 7 | 22 | 25 | 26 | 27 | 17 | 11 | 17 | 4 | 4 | 1 |
| 4′10″ | 4′11″ | 5′0″ | 5′1″ | 5′2″ | 5′3″ | 5′4″ | 5′5″ | 5′6″ | 5′7″ | 5′8″ | 5′9″ | 5′10″ | 5′11″ | 6′0″ | 6′1″ | 6′2″ |

Stature. Height is simple enough to measure and is recorded so routinely that a mass of data is available from all parts of the world.

Variation in height does not fall into neat, proscribed classes (as do blood groups) but forms a continuum of values. Nevertheless for convenience in ordinary conversation we divide the heights of individuals into such arbitrary categories as short, average and tall. If individuals are very short we talk of dwarfs or pigmies, and if exceptionally tall of giants. The range of human stature in arbitrary terms is as follows: dwarfs, very short, short, average, tall, very tall, giants, but as Figure 25 reveals the true situation is quite different. Only statistically can one find the means of describing the height classes that lie between 'short' and 'average'. This figure (a histogram) reveals the frequency of different heights within a company of soldiers: the average height is 5 ft 7 in. to 5 ft 8 in. which closely corresponds with what one would expect theoretically. The range of the sample is 5 ft 10 in. to 6 ft 2 in. The 'sample' is comprised of army recruits and if the sample had been taken from circus performers, for instance, the range would have been very much wider but the average would have been much the same.

A curious fact about stature (well known to furniture makers and garment manufacturers but to few others) is that variation in stature is principally attributable to the length of the legs. The sitting height of a group of people in a lecture theatre is remarkably uniform but when they stand up the true differences in standing height are very obvious. While this is a fairly sound generalisation for a single population, it does not hold true for mankind as a whole. Australian aborigines have relatively long legs and short bodies while mongoloids have long bodies and short legs. These racial traits recall a biological tenet called *Allen's Rule* which states, in effect, that increase in environmental temperatures tend to produce an elongation of the body, particularly the extremities. Among animals this can be seen in the compactness of tundra and arctic fauna as compared with the rangy build of tropical species. Figure 26 illustrates the relative length of the ears of American rabbits from south to north. In human populations one only has to compare the external anatomy of the Eskimo and the Nilotic African to appreciate that it is an advantage in hot regions to have a small body volume and a large surface area of skin; and that in cold regions, where heat conservation is paramount, a large volume and a small surface area is desirable.

Colour blindness. Defects of colour vision are quite common in European populations but are rare in Chinese, Japanese, African and American Indians. The condition is genetic in origin and is sex-linked; the colour defective gene is carried on the X chromosomes which are possessed by both females and males. The physical abnormality appears to be sited in the specialised, pigmented cells of the retina called 'cones', the colour-vision cells.

Colour vision defect occurs almost exclusively in males but cannot be passed from

26 Variation in ear length of N. American rabbits. *From left to right:* The Arizona Jack-Rabbit, the Cottontail of Middle N. America, and the Snowshoe of northern latitudes.

father to son although it can be passed from father to daughter who will then become a carrier. Complete colour blindness, in which reds and greens are seen as greys, is found in just over two per cent of European males. There is a less severe form of colour blindness which occurs in nearly six per cent where the ability to discriminate between red, yellow and green is affected in varying degrees. A selection of Isihara colour test charts are illustrated facing page 33. By studying these you can get some very rough idea of your own colour sensitivity.

While colour blindness is in no sense a disease it can be a considerable disability in the modern world where colours are widely used in signalling systems of various sorts, and where colour matching of fabrics, clothes, paints and so on is an integral part of our day-to-day lives. How tiresome, too, not to be able to tell whether your tomatoes are ripe or not! For most people 'tiresome' is the word to describe colour vision defects, but obviously the disability has more serious implications for some.

Biologists are naturally interested in why the deleterious genes for colour vision defects have not been eradicated from the population by natural selection. The reason is that the defects do not, in the ordinary way, affect the breeding potential of the male except inasmuch as they put a slight limitation on his choice of a career. Furthermore, there may be advantages to colour blindness that we are not aware of; we have already seen that even the sickle-cell gene, which can be a killer, bestows some advantage (resistance to malaria) on those that possess it. For example, there is one rather unexpected advantage of colour blindness – although hardly significant enough to qualify as anything but an intriguing anecdote. Individuals who are colour blind cannot be fooled by camouflage and so have a useful role to play in wartime as aerial spotters for concealed factories or gun emplacements.

Hair colour. Colour blindness is an example of a discontinuous variation but hair colour, like stature, shows an infinite gradation of colour from dark to light. Hair colour reflects the amount of a pigment called melanin present in the shaft of the hair; when melanin is completely absent the hair is white. This phenomenon which occurs naturally with increasing age may be present from birth and, in true albinos, is associated with a total lack of pigmentation elsewhere, including the iris of the eye which is pink.

Once again it is difficult to make generalisations about hair colour. We think of Scandinavians as fair, Italians as dark, Irish as red and Negroes as black; only in the case of Negroes is this anywhere near being true. Take Italians for instance; travelling from Sicily through the 'toe' of Italy northwards to the plains of Lombardy, the frequency of the swarthy, dark-haired trait falls, giving way to various shades of brown. By the time the Italian Alps are reached, blondes are extremely common. By the same score there are many dark-haired Scandinavians and a few blond Australian aborigines.

Red-heads are a bit of a mystery. The hair contains melanin plus another substance called tyrosiderin, the chemistry of which is not fully understood. The red-hair trait is definitely associated with Celtic people and so the frequency is high in East Anglia, Wales and the border counties of Scotland.

Among non-human primates, particularly the group called the colobines (leaf-eating monkeys), coat-colour changes with age. Infant colobines for the first eight to ten weeks of life have coats that are white, yellow, or bright orange in colour. As they grow older, the colour changes to the slaty greys and browns of the adults. A similar though less dramatic change is seen in several other primate species. Gorilla and chimpanzee infants show on their rumps a patch of white hair which gradually disappears with age. Adult gorilla males acquire, with maturity, a silver 'saddle' across the small of their backs. Baboon infants are totally black and only assume the olive greys and browns of the adults as they become juveniles. But of course hair colour changes in human infants too. Babies which are often black-haired at birth and golden-blond in infancy usually end up as good old mousey-browns.

One must assume that there is some selective advantage in the change of hair colour. Among non-human primates, at any rate, the evidence suggests that it is a potent identification device. The custom among colobines is for the new-born infant to be handed around among the females of the troop and the orange colour of the fur allows the mother to keep a close eye on the infant's passage from hand to hand so if danger threatens she can immediately recover her infant. This explanation does not account for the colour contrast in species that do not indulge in this so-called 'aunt' behaviour.

Hand characters. The first variable that springs to mind is of course left- or right-handedness, but handedness is primarily a matter of brain variability rather than hand variability. We are concerned with the physical characters of the hand which vary from person to person. First of all if we take the hand overall it is clear that it comes in all sorts of shapes and sizes. In the English language we have rich supplies of adjectives commonly applied to the appearance and function of the hand: beautiful, elegant, powerful, cruel. To describe the texture of the skin we can choose between rough, smooth, hairy, horny, moist, and many others. To describe function we resort to 'professional' analogies: a surgeon's hand, a musician's hand, an artist's hand, and a navvy's hand; we also talk of handshakes as being vice-like, flabby, cool, friendly, warm. Hands are things that ordinary people notice and on which they put great reliance as indicators of character. Like all generalisations, this sort of typological thinking can lead one into serious errors of judgement. The most brilliant plastic surgeon that I know (and no profession demands greater delicacy than plastic surgery) has a short, powerful 'square' type of hand with stubby fingers that would have been the ideal model for the power-oriented fists of our caveman ancestors.

Apart from the collective character differences that give human hands their individuality, there are a number of indivi-

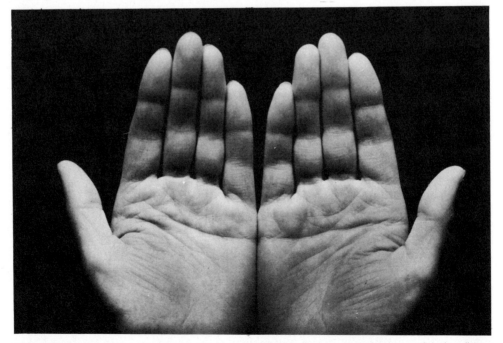

27 The left hand of female subject shows a 'heart' line broken opposite base of index finger (common human pattern). On the right this crease extends right across the palm (ape and monkey pattern). Note short index finger.

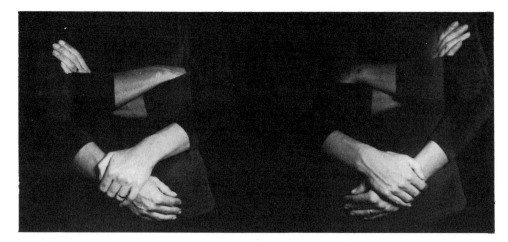

28
Arm-folding
and wrist-
clasping
(see text).

dual characteristics that anthropologists use as markers to establish genetic variability within a species:

Hand clasping. Clasp your hands together and note which thumb lies on top; or fold your arms and note whether the right or left wrist is uppermost. Now try it on your family and friends. For arm-folding, the ratio is 6:4 in favour of the left wrist, and for hand clasping 6:5 in favour of the right thumb. It would be interesting to determine whether proportions differ between sexes.

Length of index finger. Hold up your hands, palms towards your face, and note the relative projection of the ring and index fingers. If your index finger is longer than your ring finger you are probably a female, if shorter then the probability is that you are male. Whether you are male or female a long index finger should boost your self esteem because you possess a progressive human trait. If the reverse is true, don't despair but take comfort from the fact that your particular pattern is hallowed by time. You possess the pattern of your primate ancestors and the genes involved are at least sixty million years old. You may find, like me, that you are 'progressive' on one hand, and 'primitive' on the other. What does that make us? Ape-men perhaps?

29
Hand-
clasping
in two
different
subjects.
What do
you do?

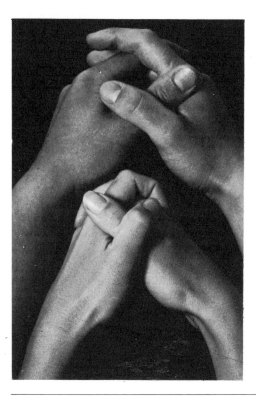

'Lines of fate'. The lines and creases of the palm are the bread-and-butter of the palmist and the fortune-teller, but for the anatomist they are signs of man's material-istic present, not portents of his unknown future. The lines of the palm are simply flexion creases – skin-hinges. Look at your hand and bend your fingers and you will see what I mean. If you are wearing a coat, a sweater or a cardigan as you read this, look at the creases that appear in the material as you bend your elbow.

There are a host of minor variables in the detailed anatomy of palmar creases, which apparently mean much to palmists but are totally irrelevant to the rest of us who have, by now, absorbed the idea that no human beings are quite the same and that this variability has no mystical significance. A major variable which can be seen at a glance is the well-marked crease nearest to the base of the fingers (the 'heart line' of palmistry). In most of us this crease starts opposite the base of the middle finger. Here there is a break, and a second – less well-marked – crease, starting in the middle of the palm, takes over and forms the skin-hinge for the index finger (Figure 27). The functional explanation for this arrangement is quite easy to understand but is irrelevant to the present discussion. In a small proportion of individuals there is no 'break' and the crease runs from one side of the palm to the other. This sort of crease is typical of monkeys and apes, and if you happen to be the possessor of such a crease you need not read anything sinister into the fact – it doesn't make you less human! On the contrary, you are a human being who carries on his hand a little bit of the evidence which proves the validity of Darwin's theory of evolution.

PTC tasting. So far we have been concerned largely with characters that can be observed with the naked eye or revealed by simple visual tests but there are a large number of biochemical variations that can be determined from urine analysis, blood analysis and so on. There is, however, one variable which can be demonstrated without recourse to complex methodology. The ability to taste the compound phenylthiocarbamide (PTC) is known to be a genetically controlled trait. A drop of PTC in very low concentration (1/20gm in one litre of water) placed on the tongue produces one of two reactions: either the recipient grimaces violently and expresses all the emotions associated with a revolting taste, or he smiles blandly and asks what the fuss is all about. Humans are 'tasters' or 'non-tasters' (so, incidentally, are non-human primates). Tasters outnumber non-tasters by 4:1.

There is no simple rationale for PTC tasting. It is not a substance that is usually found in food. But, nevertheless, it points the way. There are people who cannot bear the taste of this or that. There are curry addicts and those who loathe the taste; onions have their fans and garlic its implacable enemies. Green or red peppers, olives, crabs and lobsters, smoked salmon, caviare, mangoes, guavas, avocados are either acceptable or non-acceptable. We know virtually nothing about the roots of human gustatory preference. It is highly likely that they have a genetic basis (like PTC) but we are far from understanding the mechanisms.

I think I have cited enough examples of natural variation in man to show that there is, indeed, 'nowt so queer as folk'. Folk are queerer than animals (who are queer enough, heaven knows) simply because their genetic stockpile is larger and so provides the potential for a greater number of genetic variations.

Animal populations are limited by geographical and climatic factors from the consequences of crossbreeding between foreign members of the same species; they tend to interbreed in a restricted local sense just as the inhabitants of rural England used to do before the invention of the bicycle which permitted courting village swains to disseminate their genes over a wider area than ever before. Modern man is a universal species and the world is his oyster by permission of BA and TWA and all the other airlines that transport people and their genes from one part of the globe to another. His promiscuity (in the nicest sense of the word) is the cause of his queerness (also in the nicest sense) and the ultimate guarantee of his survival as a species.

30
Variation in dogs results from selective breeding master-minded by man for his own advantage and is not comparable to natural variation which has been the subject of this chapter.

31 In the beginning . . . an infant primate, a legatee of sixty-five million years of environmental change.

3

In the beginning . . .

Man is an animal. No doubt, this uncompromising statement still comes as something of a shock to some but not, I suspect, to many. We have had over a hundred years to absorb the idea of man's animal heritage and even such rabid critics of Darwinism as Bishop Wilberforce, Richard Owen – the great naturalist – and Benjamin Disraeli would surely, as rational men, have by now accepted the truth of the unpalatable fact. While undoubtedly there are many vociferous pockets of resistance, principally to be found among fringe religious sects in the 'bible belt' of the United States (and, surprisingly, a strong anti-evolutionary body exists in the state of California), the orthodox religions accept the fact of evolution, albeit with varying degrees of enthusiasm. In the light of the liberal attitudes of the second half of the twentieth century, with its overtones of a back-to-nature movement (its critics would call it permissiveness), the present-day reply to Disraeli's rhetorical question, 'Is man an ape or an angel?', would be overwhelmingly in favour of the ape.

MAN'S PLACE IN NATURE

The evidence for man's animality is overwhelming. Developmentally, physically, physiologically and even behaviourally he is in immaculate accord with the corresponding characteristics of all animals from the lowliest to the highest. Man is just as much a part of nature as a beetle, a snake, a hippopotamus or a chimpanzee. Zoologically he offers no problems to classifiers; man is in the *Class Mammalia*, *Order* Primates, *Suborder* Anthropoidea, *Superfamily* Hominoidea, *Family* Hominidae. Taxonomically he is a fully paid-up member of the trades union of the animals of the world.

There is no doubt that man is different from the rest but this difference is really no greater than that which separates aquatic mammals from land mammals. Man has broken through a behavioural barrier (just as whales and porpoises have broken through a water barrier) and entered a world where the environment is no longer natural. The components of human culture – language, art, religion, sociality and technology – provide the ingredients of this new ecology; the natural forces which determined the course of the old ecology have little relevance for man. However delightful it may be in theory, there can be no return to nature. Nature for man is no longer natural.

However, there is another supremely important force acting on human beings, and this is the matter of genetic inheritance. Man's basic anatomy and physiology and many of his likes and dislikes, actions, emotions, instincts and drives owe more to the past than they do to the present. We are faced with something of a paradox; here we have a creature with an ancient animal body and ancient animal instincts trying to survive in a modern environment of his own creating for which he is only imperfectly fitted. It is not surprising that he is in perpetual conflict within himself. He is being pulled into two different directions by the impelling memories of 'nature' on the one hand and day-to-day directives of 'nurture' on the other.

What then is man's place in nature? He is both inside it and outside it. He is an animal but he is also a man. He cannot return to nature, for nature is no longer natural, but equally he cannot escape the grip that nature has on his basic systems and instincts. No wonder he has problems; no wonder that the urge for the simple life and an unpolluted atmosphere keeps recurring in contemporary society. Jean-Jacques Rousseau and his followers in the eighteenth century with their 'noble savage' philosophy and their conviction of the beauty and innocence of nature, set the fuse (as Lord Clark put it in *Civilisation*) on a 'time-bomb which after sizzling away for almost two hundred years has only just gone off'.

The epithet *man, the peculiar animal* which was coined by an old friend and colleague, Richard Harrison, is a very appropriate description of our crazy, mixed-up constitution. One might say that man's principal peculiarity is his overwhelming concern with just how peculiar he is.

With man, the peculiar animal, in some kind of perspective we can turn our attention to the long journey of the mammals and the primates. Their story is our story.

EMERGENCE OF THE MAMMALS

In the beginning there were just two kinds of mammals, separate descendants of early offshoots of the reptiles which appeared in the fossil record 150-170 million years ago during the period known as the Jurassic. One stock gave rise to the monotremes (egg-laying mammals, e.g. duck-billed platypus) and the other to the marsupials (pouched mammals, e.g. kangaroo, koala bear, etc.) and placentals (advanced mammals, e.g. primates, giraffes, etc.). Today their distribution is as shown in the table at the foot of the page.

In terms of both numbers of animals and numbers of species, the placentals far outweigh the marsupials, and lagging a long way behind are the egg-laying monotremes. There are only two genera of monotremes, a once widely distributed infraclass: the duck-billed platypus and the echidna or spiny anteater. The marsupials seem to be nearer our own line of evolution than the monotremes, which may have evolved independently, and so can probably be regarded as the earliest of our kind of mammal.

As the primates, which are our principal concern, are eutherian mammals – placentals – regretfully, we shall have to forgo the pleasure of delving more deeply into the life history of the prototherians and metatherians which include the kangaroos and wallabies, the 'rabbits' of Australia, and everyone's favourite – the koala bear.

The ancestral placental mammals were fairly small, quadrupedal creatures with short limbs, shortish tails, five-fingered hands and feet, long snouts and a full set of teeth (11 in each half of each jaw: 3 incisors, 1 canine, 4 premolars, 3 molars, which were sharp and pointed; from the dental character it can be inferred that their diet was largely insectivorous. As most living insectivores are solitary animals which are nocturnal in habit

CLASS	SUBCLASS	INFRACLASS	COMMON NAME	GEOGRAPHIC DISTRIBUTION
Mammalia	Theria	—	Mammals	World-wide
Mammalia	Theria	Prototheria	Monotremes	Australia and New Guinea
Mammalia	Theria	Metatheria	Marsupials	Australia, New Guinea, North and South America
Mammalia	Theria	Eutheria	Placentals	World-wide except Australia, New Guinea, New Zealand

and do not form complex social groups, the tendency, rightly or wrongly, is to assume that their ancestors shared this behaviour.

The earliest placental mammals belonged to the Order Insectivora and were living in an environment which was very different from that of today. The climate was hot and humid and the low-lying land surface was broken up by lakes, swamps and inland seas. The vegetation was dominated by the gymnosperms such as conifers, cycads and various species of giant ferns. Angiosperms in the shape of flowering shrubs, like the magnolia, were beginning to appear. The most successful land animals at this time (Cretaceous) were the dinosaurs. Some of the colossal quadrupedal, plant-eating dinosaurs like *Diplodocus*, *Brachiosaurus* and *Stegosaurus* had already vanished, but their successors though not quite so large were infinitely more menacing. The largest flesh-eater of the time was the horrific *Tyrannosaurus rex*, twenty foot high and fifty foot long. *Tyrannosaurus* was the biggest land predator the world has ever known. Like many cretaceous reptiles it walked on two feet and the thunder of its passage must have put fear into the heart of many a small and defenceless mammal (Figure 32). But the rule of the dinosaurs was nearly over. At the

32 The Dinosaurs, *Triceratops* and *Tyrannosaurus rex*. Note small mammal making a hasty exit on left.

end of the cretaceous period some very unpleasant things started to happen – unpleasant for the dinosaurs that is.

The mention of Laramie means different things to different people. For addicts of TV westerns it means many hours of pleasurable entertainment, but to geologists it recalls the tectonic Armageddon known as the Laramide revolution that put an end to the Middle Period of the earth's history and explosively ushered in the new. For both TV addicts and geologists Laramie has strong dramatic overtones, but it took Walt Disney's genius to blend the essential theatricality of the event with its academic significance. In a wonderfully imaginative sequence the Laramide revolution was re-created by Disney in 'Fantasia' to the music of Stravinsky's 'The Rite of Spring'. In this sequence one saw the last of the dinosaurs battling for survival against the heat and desiccation of the final breathless and sun-searing moments of the Cretaceous period. Earthquakes split the ground wide open, volcanoes belched forth molten rock, mountain ranges erupted, rocks cracked and toppled, immuring the ruling reptiles, like the inhabitants of Pompeii, under the laval flows of Vesuvius, in permanent stony sarcophagi. Unscientific though this sequence may have been, its impact was unforgettable.

The effects of the Laramide revolution, which lasted for several millions of years, left their mark on the shape of the land surface, the climate and the vegetational cover. Inevitably the balance of animal life was disturbed. The principal victims were the dinosaurs which were swept away in what seems to us, at this remote distance of time, as the twinkling of an eye. The beneficiaries were the mammals, the insects and the birds which grew into greatness under this new regime.

The sudden extinction of the dinosaurs was probably not as 'sudden' as all that. The process of diminution of known reptilian families proceeded steadily through the late part of the Cretaceous period. When two prolific dinosaur sites in North America, separated by several hundreds of thousands of years, are compared, the numbers of genera of dinosaurs present falls from 81 to 38, the biggest casualties being among the plant-eating dinosaurs. Nevertheless, the cutoff point when it came was dramatic in its speed and completeness as the following description attests:

'In Utah there is a sequence of sediments known as the North Horn formation. This is a continuous series of sandstones and clays without any visible break, without any evidence of disturbances that might have interfered with the steady accumulation of muds and sands year in and year out. By all ordinary geological criteria the North Horn formation seems to show a single cycle of sedimentation yet in its lower part are the bones of dinosaurs and in its upper part the bones of mammals. There is no extension of the dinosaurs up into the mammal zone. The separation between the uppermost dinosaurs and the lowermost mammals is no more than thirty-five feet, surely not a great hiatus in the geological sense . . . Here we see the physical evidence for the short time span during which the ruling reptiles vanished from the earth. Certainly it was a sudden event in geologic terms in ancient Utah, and so in general it would seem to have been all over the world.' (*The Age of Reptiles*, p. 202.)

It is not easy to visualise the sort of happenings that could produce such a great extinction, in fact it is still one of the major unsolved mysteries of zoology. Disney attributed it to the earth movements and the climatic and vegetational changes accompanying the Laramide revolution but this we know was no 'overnight' affair. It could hardly account for the final, sudden extinction although it is most likely to have contributed to the preceding decline.

The obvious explanation is some sort of acute global catastrophe – a flood or something of the sort, but if such an event took place it is not written in the rocks of Utah or anywhere else in the world. Furthermore it would be an odd catastrophe that was so zoologically selective that it killed off some reptiles but allowed other reptiles and other groups of animals, including mammals, to survive. The next possibility is a catastrophe of extra-terrestrial origin, an exploding star which drenched the earth with abnormally high doses of cosmic rays, a galactic explosion in the Milky Way or a collision with a giant comet. Once again the rocks are silent, and the zoological evidence is lacking. Other suggestions have been made: perhaps the explanation might be in the activities of the mammalian underground which attacked the dinosaurs at grassroots level by the subtle ploy of preying on their eggs. Eggs of birds and reptiles are of course a constant target of mammalian predators but reptiles lay eggs in such vast numbers that statistically the chances of eradicating a large population by this means are very low indeed. Anyway, what about marine reptiles that vanished at the same time? They are believed to have given birth to live young, not eggs.

Whatever the reason, at the end of the Cretaceous the Age of Reptiles gave way suddenly to the Age of Mammals. The subsequent Cenozoic era was not only the Age of Mammals but it was also the Age of Angiosperms (flowering shrubs). The monochromatic landscape of the Mesozoic dominated by the greens of conifers, ferns and waterplants was succeeded by a greater variety of flower- and fruit-bearing plants that not only coloured the scene and attracted the insects, but turned the forests into attractive places for animals to live. Trees became for the first time desirable residences and at the head of the queue for occupancy were the primates.

The first placental mammals to emerge as a recognisable Order were the insectivores but soon they had diversified into three major groups: the primates, the creodonts – flesh-eating mammals, which ultimately became extinct – and the condylarths, the ancestral ungulates or hoofed mammals. These groups are now quite distinctive. One does not have to possess much special knowledge to pick out a mole, a monkey, a tiger or a cow, but in the beginning this would have been a major problem. Supposing you could step into a time machine and set the space-time coordinates for a forest in North America early in the Palaeocene. You would find yourself in a rich subtropical forest surrounded by tall familiar trees like poplars,

willows and oaks arranged in the two- or three-layered patterns that can be seen today in forests all over the world. So familiar would be the scene that your first reaction might be to wonder whether you had set your coordinates correctly. Then you would see the animals and you would know that you had made no mistakes; they would be vaguely familiar but at the same time disturbingly odd. Finally the odd thing would dawn on you; whether they were scurrying along the branches of trees, fossicking among the debris of the forest floor, or emerging, blink-eyed, from burrows in the ground, they would look very much alike. Knowing what you know about mammalian evolution you would be hard put to it to say, 'Ah! That's a primate . . .', 'That's an ancestral carnivore', or 'That's obviously an evolving horse'.

Had you returned to the same forest a few million years later you would have seen a distinct change; evolution had been proceeding among the mammals at a cracking pace. Insectivores and primates were erupting in bewildering number and variety, and new orders had appeared. As new orders evolved, a great deal of readjustment had taken place between animals and the life-zones within the forest. The primates had taken to the trees and were already characteristically adapted to this new environment. You would be able to recognise lemur-like creatures leaping from branch to branch or clinging to the vertical trunks of saplings. On the ground, dog-like animals like *Hyaenodon*, a sort of primitive hyaena, and *Oxyaena*, an early cat-like carnivore, were stalking their prey, just as leopards and tigers do in tropical forests today. Lumbering about feeding on the foliage were giant hippopotamus-like amblypods (*Uintatherium*), and if you were lucky you might catch a glimpse of a 'dawn horse', the tiny, shy, twelve-inch-high *Hyracotherium* (= *Eohippus*).

At this point in time – the Eocene epoch some fifty-five million years ago – we are going to change the lens which is the eye of our time machine. Up to now we have been using a wide-angle lens to embrace a broad spectrum of animal life; now we need to switch to the telephoto and zoom in on a relatively small area of the mammalian scene in order to see in more detail how our remote ancestors, the primates, were coming along and how they were acquiring the physical skills and mental talents that have subsequently stood mankind in such good stead. But first there is the pressing matter of time . . .

TIME, TIME-CLOCKS AND CALENDARS

Evolution has been defined as 'change with time'. Already in chapter 2 we have considered the causes of change but have done little about time except to juggle with geological periods, eras and epochs as if they were Indian clubs in a circus act. So it's about time that we took a brief look at time.

Ancient time is usually referred to as geological time simply because it is only through the study of the rocks laid down long before animal life appeared on this planet that accurate estimates of the age of the earth can be calculated. The basic method is known as radioactive dating, a technique which depends on the behaviour of certain elements like uranium, rubidium, potassium and carbon. When the atomic particles of these elements 'decay' or break down they change at a predictable rate into stable substances such as lead, strontium or argon gas. Thus, given a known speed of decay, it is relatively easy to estimate the age of a sample of rock by measuring the proportion, for example, of uranium/lead or rubidium/strontium. The oldest rocks which have been dated by the radioactive method are found in Greenland at the mouth of the Ameralik fjord; they are said to be between 3700 and 3900 million years old.

Faunal dating, dating by fossil, remains of animals, is a less precise method inasmuch as it cannot provide absolute dates. Faunal dating can only help to arrange rock systems into sequences on the basis of their contained fossils. As personal judgements of identification are involved, a subjective element is present. Isotopic dating using radioactive materials is founded on firmer, more objective grounds which can, at any time, be confirmed by experiment. Faunal dating, however, occupies an important place in determining the chronology of the past because by the comparison of fossils found in different parts of the world, correlations between widely separated rock formations can be made, and a global picture can be established.

In practice, chronological tables are derived from both sorts of dating methods. Geology has given a time-scale for the zoological succession. For example in the table below the first column lists the three main

Era	Period	Duration (in millions of years)	Time-scale (in millions of years)	Dominant form of life
CENOZOIC	QUATERNARY	2	2	MAMMALS
	TERTIARY	63	65	
MESOZOIC	CRETACEOUS	71	136	
	JURASSIC	54	190	
	TRIASSIC	35	225	REPTILES
PALAEOZOIC	PERMIAN	55	280	
	CARBONIFEROUS	65	345	AMPHIBIANS
	DEVONIAN	50	395	
	SILURIAN	35	430	FISHES
	ORDOVICIAN	70	500	INVERTEBRATES
	CAMBRIAN	70	570	

TIME-SCALE FOR THE HISTORY OF ANIMAL LIFE

Ancient Environments: 33 EOCENE Small ungulates and a dog-sized carnivore (*Hyaenodon*) can be seen. In the tree is *Pelycodus* (an early primate) and lower down a little marsupial called *Peratherium*. Vegetation still shows the conifers, cycads and ferns of the Cretaceous.

34 OLIGOCENE In the foreground is a large, horse-like chalicothere which had claws instead of hooves on its feet. There are also hyaenodons and an early rhinoceros.

eras which are expressed in zoological terms: *Palaeozoic* meaning early forms of life, *Mesozoic* and *Cenozoic* meaning intermediate and recent forms of life respectively. The second column is couched in geological terms; periods are arranged according to the relative ages of rock formations.

The era with which we are mainly concerned is called the Cenozoic and in the table the subdivisions, or epochs, of this era are shown with their approximate durations in millions of years. Once again, epochs are geological concepts.

Quaternary	PLEISTOCENE	2
	PLIOCENE	3
	MIOCENE	18
Tertiary	OLIGOCENE	13
	EOCENE	18
	PALAEOCENE	11
	TOTAL	65

Cenozoic time-scale showing duration of each epoch in millions of years

It is generally agreed that the fossil record of animals goes back 600 million years, approximately one-eighth of the age of the earth which is estimated at 5000 million years. The length of time that man – our sort of man, *Homo sapiens* – has been around is about 250,000 years or 1/20,000 of the time since our planet first fell out of the sun.

If these sort of figures don't throw you then perhaps you are (or ought to be) a mathematician or an astronomer. But to most of us, time expressed in millions of years is as incomprehensible as space expressed in terms of light-years. Although it is strictly designed for us simple-minded people, I think it is helpful to convert the millions of years of geological time into the more familiar, more digestible scope of a calendar year. Let us suppose that we decide that the 600 million years of the fossil record is equivalent to 365 days. When we do this we

find that events occurring in the inconceivable past are painlessly converted into the happenings of a few months, weeks or days ago. For example if we are told that vertebrates, animals with backbones, have been in existence since 23 March and that primates originated on, or about, 22 November, we are in a much better position to appreciate the tempo of the evolutionary process than if we had been told that 540 million years separated these two events.

The analogy becomes very dramatic when human evolution is involved. For instance it was not until 24 December that the first man-like creature appeared in the fossil record. We have to wait until 30 December before the first representative of our own genus *Homo* walks into the arena. Now comes the cliffhanger: New Year's Eve dawns and there is no sign yet of *Homo sapiens*. The day wears on . . . 5 o'clock, 6 o'clock, 7 o'clock and still no sign! Eight o'clock comes and the minutes tick by . . . 11 . . . 12 . . . 13 . . . 14 past the hour and then . . . suddenly . . . at 8.15 pm on the last day of the year, in walks *Homo sapiens*. Whether he was tall and dark or blond and squat is really immaterial but he was certainly a stranger.

The calendar analogy of course takes many liberties with accuracy but no more than does the formal geological time-scale. The fundamental source of error lies in the dating technique itself. No known method is more than seventy-five per cent accurate. Thus with a period of ten million years the estimates could be 'out' by as much as 2.5 million years. This explains the airy insouciance of the palaeontologist who will remark that this or that fossil is 20 million years old give or take a few million! The dates on our calendar are based on the most commonly accepted dates for these events.

CONTINENTAL DRIFT

The Cenozoic was a period of environmental change. It started with a revolution, and revolutions of one sort or another characterised the 65 million years of its duration. There was the Cascadian revolution of

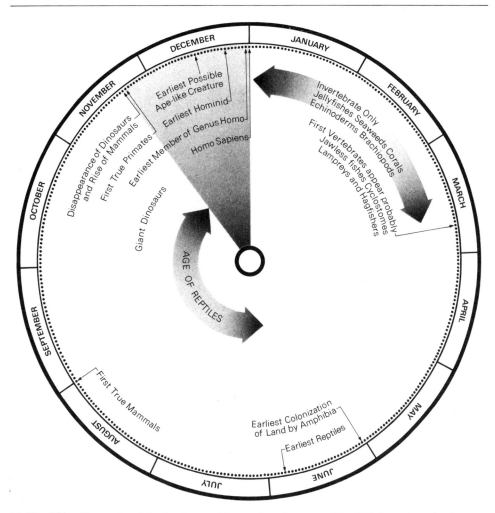

The following labels appear on the calendar diagram:

- DECEMBER
- JANUARY
- NOVEMBER
- FEBRUARY
- OCTOBER
- MARCH
- SEPTEMBER
- APRIL
- AUGUST
- MAY
- JULY
- JUNE

- Earliest Possible Ape-like Creature
- Earliest Hominid
- Earliest Member of Genus Homo
- Homo Sapiens
- Disappearance of Dinosaurs and Rise of Mammals
- First True Primates
- Giant Dinosaurs
- First True Mammals
- AGE OF REPTILES
- Earliest Colonization of Land by Amphibia
- Earliest Reptiles
- Invertebrate Only Jellyfishes Seaweeds Corals Echinoderms Brachiopods
- First Vertebrates appear probably Jawless fishes Cyclostomes Lampreys and Hagfishers

35 The 600 million years of the fossil record is seen here in terms of the 365 days of a calendar year.

Miocene times and, during the Pleistocene, there were the revolutions of the ice ages which became as regular a feature of the epoch as palace revolutions in some South American republics today. The Laramide revolution and the Cascadian revolution can probably be regarded as world-wide events associated with the phenomenon of continental drift, but the ice ages are a rather different proposition. Many suggestions have been made to account for the ice ages and Nigel Calder discusses some of them – both reasonable and outlandish – in his book *The Restless Earth*. He is forced to conclude that, like the sudden demise of the last of the dinosaurs, the phenomenon of the Pleistocene ice-ages are still a total enigma.

The theory of continental drift has had an interesting history. It was first advanced as a serious scientific proposition by Alfred Wegener, a German meteorologist, in 1912. There followed a long – too long – interregnum of disbelief and disinterest when zoologists were happier to invent lost continents or vast transoceanic land 'bridges' to account for the movement of animals than to delve into the inevitable logic of Wegener's ideas. In the last ten years or so the whole attitude of science has changed. Research in rock-magnetism, seismology and particularly sea-bed geology has so compellingly argued the case for continental drift that it has become an article of faith among earth-scientists. As Nigel Calder remarks, ten years ago a continental-drifter wouldn't have stood a chance of being appointed to a reputable university post in the United States, but nowadays he is damned from the start if he is not a believer. To doubt continental drift is almost a cardinal sin; you might as well be a flat-earther if you deny that continents have been floating around like rafts on a sea of magma for the last 250 million years. Personally I am a continental-drifter but I am not yet prepared to explain every botanical event, every zoological puzzle in terms of wandering continents.

The theory is that at the end of the Jurassic all the present continents of the earth were still united in a single land mass called *Pangaea* (Figure 36a). All parts of this supercontinent were in contact so, theoretically, the land animals then living were common to all parts. Towards the end of the Jurassic period, *Pangaea* began to break up. The first continents to drift away were Antarctica and Australia (including New Guinea and New Zealand). Peninsular India also broke away from Africa at this time and a rift appeared between N. and S. America (Figure 36b). The rest of the supercontinent became all but completely separated into Laurasia in the north (Europe, Asia, North America) and Gondwanaland in the south (South America, Africa). It is believed that the marsupials had evolved before Australia and Antarctica started to drift away. If this hypothesis is correct it would explain how only monotremes and marsupials come to be found in Australia; and why no eutherian, or placental, mammals of any sort are to be found there except those that have subsequently been introduced by man. If we assume that the placentals evolved after the supercontinent had started to fragment and drift apart, it is clear why they missed the boat to Australia. There must have been many marsupials which were left behind, but in competition with the more efficient placental mammals they were eventually all but wiped out. Only two families remained – on the South American continent.

The family Didelphidae are represented in North and South America by many species of opossums. In addition in South America there are a few species of shrewlike marsupials belonging to the family Caenolestidae. No doubt their survival is linked with the rather peculiar zoogeographical history of South America which was isolated from migrations from the north for the best part of 50 million years before the isthmus of Panama finally assumed its present form towards the end of the Pliocene. Certainly some rodents and primates managed to bridge the gap in the late Eocene by the chance methods of island-hopping (see page 70). By the time the Isthmus of Panama was formed at the end of the Pliocene and the major invasion of recent mammals took place, these American marsupials were in no danger of being displaced from their well-established ecological niches.

It is assumed that on most of the supercontinent of *Pangaea*, marsupials were replaced by placental mammals during middle to late Cretaceous times. Placental mammals can be recognised in the fossil record from the late Cretaceous onwards.

During the Cretaceous a further break-up of *Pangaea* had occurred (Figure 36c); South America had started its westward drift and the South Atlantic was beginning to open up along a north/south axis which, today, is represented by the site of the Atlantic mid-ocean ridge, the ancient junction between two huge continental plates (South America and Africa) which gradually slid further and further apart. By the late Cretaceous the South Atlantic was some 1875 miles (3000 km) wide, indicating a speed of drifting of 1.6 inches (4 cm) per year.

At the start of the Cenozoic era, at the beginning of our story in fact, North America was still joined to Eurasia by two land bridges. At the west end was the Atlantic bridge consisting of what is now Labrador, Greenland and the British Isles, soon to be opened up. At the east end was the Bering Strait land bridge linking eastern Alaska with Siberia. Europe was intermittently linked with Africa across the Tethys Sea, the open waterway to the east of the Straits of Gibraltar (Figure 36d) that was then continuous with what is now known as the Arabian Sea and the Indian Ocean. North and South America were separated by a wide channel, and peninsular India was still drifting in a north-easterly direction towards its ultimate Asian landfall along a line known

36 Continental Drift:
Dotted lines indicate present-day continental coastlines; shaded areas indicate shallow epicontinental seas.

a The World in the late Jurassic, 140 million years ago

b Mid-Cretaceous, 105 million years ago

c Late Cretaceous, 75 million years ago

d Upper Eocene, 50 million years ago

e Miocene, 20 million years ago

f The World today

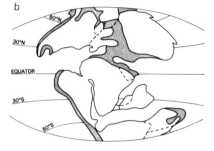

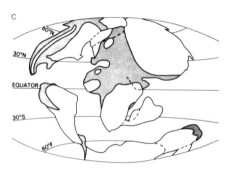

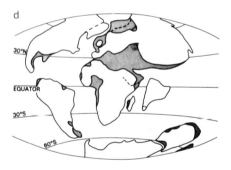

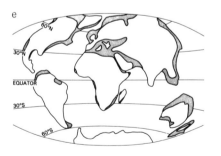

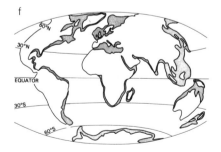

as the Himalayan 'suture'. The archipelago of South-east Asia was as yet under the sea, as were probably most of the existing Pacific islands. Finally, Madagascar, the home of the lemurs, was already an off-shore island of the African continent.

The reason why continental drift has suddenly become a respectable theory is that scientists have now come forward with a convincing explanation as to how it happened. There is an analogy here with Darwin's theory of evolution. This great idea, based on the interpretation of the evidence that Darwin saw all around him, had its adherents who believed in it intuitively and emotionally without any very good reasons for doing so. However scientists as a whole, who are rather a hard-headed lot, were downright sceptical. But in the early years of the twentieth century the work of Gregor Mendel, the Austrian monk, on the mechanisms of inheritance was rediscovered and a whole new subject of genetics was born. It was then that scientists, comforted by the assurances of the experimental method, started to take Darwin's theory seriously.

So it is with continental drift. The turning point was a concept called plate tectonics that seemed to synthesise and unify a number of independent observations. The theory of plate tectonics supposes that the outer shell of the earth, some forty miles in depth, is composed of a series of rigid, unconnected plates floating on a 'sea' of molten rock or magma. As the convection currents in the magma swirl with a circular motion from within out, they set in motion the continental rafts floating on the surface. When these rafts, or plates as they are properly known, come up against other plates the pressure exerted squeezes up the rocks to form mountain ranges. The impact between the South American plate and the Pacific plate for instance threw up the Andes mountains; and the crunch of the wandering Indian plate when it made its landfall with Asia gave rise to the Himalayas.

Separation of plates, on the other hand, leads to a welling-up of the semi-fluid magma from the mantle layer of the earth to fill in the gap in the crust; this produces the mid-ocean ridges at sea and rift-valleys on land. Plates can also override one another, the overridden plate being forced down at a steep angle into the mantle layer leaving a depression on the surface known as an ocean trench. It can be shown that major changes in the contours of the earth's surface occur only at the boundaries of plates; these boundaries correspond closely with the earthquake zones of the world.

While many different observations led to the formulation of the synthetic theory of plate tectonics, the contribution of palaeomagnetic studies has been outstanding. Put simply, certain rocks contain iron-bearing minerals whose particles act as compass needles and point towards the world's magnetic poles. When the directions in which these particles point are plotted from widely separated geographical areas, it appears that in times past the poles have been wandering all over the globe. This is regarded as highly unlikely – far less likely in fact than the alternative explanation which is that *it is the continents themselves that have been moving*, both relative to the poles and relative to each other (Figure 36).

One interpretation of the palaeomagnetic data is that 400 million years ago *Pangaea* was centred on the South Pole with Europe, Asia and North America on the near side of it and Africa and South America on the far side. Since then the whole mass has been sliding round the bottom of the world and drifting northwards. By the period known as the Carboniferous, when the tropical forests of Europe and North America provided the basis of the coal-bearing strata of today, these two continents lay approximately on the Equator. The northward shift of the land masses continued right through the Tertiary period. During the same period there was a steady southward movement of tropical mammals such as the primates from Eurasia and North America to Africa and South America. In terms of geography, southward movement meant migration from one continent to another but in terms of latitude their position was unchanged. Like

37 MIOCENE The reconstruction includes an active volcano, reflecting the volcanism and mountain-building that characterised this epoch. Mammals are beginning to assume a rather modern form, chevrotains, rhino and elephants can be seen. In the tree on the right there are ape-like primates, probably of the *Proconsul* group.

the Red Queen in *Through the Looking-glass*, they were running in order to stay in the same place.

While continental drift is obviously the primary factor influencing the present distribution of primates, there are a number of secondary factors. Some of these are them-selves consequences of continental drift, but others may require a different sort of explanation. In the following section we shall be considering the effects on primate zoogeography of changing climates, altera-tions in vegetational patterns and fluctuations in sea-levels.

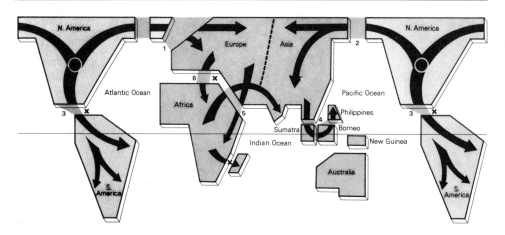

38 Possible migration routes of primates during the Cenozoic (tertiary period). Centre of dispersal is postulated as North America. Numbers indicate existence of land bridges at various times. (1) North Atlantic; (2) Bering Strait; (3) Central America; (4) Sunda Shelf (East Indies); (5) Afro-Asian; (6) Possible site of Afro-European bridge. X indicates where 'Rafting' may have taken place.

CLIMATE, VEGETATION AND LAND BRIDGES

Since the beginning of the Cenozoic, world temperatures have been steadily dropping (Figure 39). During the Eocene the mean annual temperatures in Seattle, Paris and London were comparable to those experienced today in Mexico City. Subtropical forests extended as far north as 53° latitude. At the end of the Eocene, the New Siberian Islands well within the Arctic Circle enjoyed a temperate climate as evidenced by the nature of the fossil plants. During the Oligocene there was a sharp acceleration in the cooling process: subtropical forests reached no further north than Los Angeles. In western Europe warm temperate forests had replaced tropical forests and yearly mean temperatures were in the neighbourhood of 64°F (18°C). In the Miocene the climate of California was warmer than it is today, but it was not subtropical. By the upper Miocene, average temperatures of the world were much as they are at present but the climate was more equable, there being less contrast between summer and winter. By the Pliocene, permanent polar icecaps had formed and arctic weather was moving southwards, bringing conditions for glacier formation into the middle latitudes – heavy winter snowfalls and cool, cloudy summers.

During the latter half of the Pleistocene the polar ice extended far southwards and brought arctic conditions into regions of the Northern Hemisphere that are now temperate. The ice-sheets, unable to melt in the cooler summers, crept down from the north pre-empting unaccountable tons of water which appreciably lowered sea levels throughout the world and raised previously submerged continental shelves linking Iceland to Britain, Britain to Europe, and Europe to Africa. Tundra conditions, associated with mammals of cold regions such as the musk ox and walrus, extended into California, Texas and Mississippi.

Between the four great Ice Ages – the Gunz, the Mindel, the Riss, and the Würm – were long intervals or 'interglacials' when warmer conditions prevailed and when mammals, including non-human primates and man, swarmed back into the then temperate regions of Europe, Asia, and North America from whence they had been driven by the encroaching ice-sheets.

The effect of these climatic changes on the distribution of primates was considerable. In the Eocene, subtropical and tropical forests spread 50° north and south of the equator providing a wide habitable belt of 100° latitude; today the belt of subtropical and tropical vegetation – the home of primates – has shrunk to less than 50° latitude. Many of the areas between the tropics of Cancer and Capricorn are either deserts, grasslands, mountain regions, or high plateaux. There is very little forest. The primate world has shrunk since the days when primates abounded in the vast forests of North America, Europe, Asia, and southern England. The actual area of forest habitat presently available is probably less than five per cent of what it used to be sixty million years ago.

It is important in the study of primate taxonomy and evolution to know something about the migrations of primates during the Cenozoic and to determine when and by what routes they reached the areas in which they are now found. In Figure 38 the solid black spots represent the centre of dispersal and the arrows represent the routes; the numbered spots indicate land bridges or sea routes which have been open at various times in the last sixty-five million years. It can be seen that, in the absence of other evidence, North America is provisionally regarded as the birthplace of primates. From North America they reached Europe and Asia by two routes: via the North Atlantic bridge to Europe and via the Bering Straits to Asia. Primate migrations into Europe and Asia could only have occurred in the early part of the Tertiary when the Asian and European land bridges were still warm and forested to permit the passage of primates used to tropical conditions. It is unlikely that there were any further movements of primate stocks between Asia, Europe, and North America after the middle Eocene.

From North America, primates reached Central America via a land connection, but the passage to South America was blocked by a broad seaway which separated the two continents from the middle of the Eocene until the Pliocene. Nevertheless, it is certain that a crossing did take place as fossil primates are found in South America dating from the Oligocene. The likely mechanism is by a process of 'rafting' which implies a chance crossing of rivers or open water by means of floating vegetation, tree trunks or blocks of matted grass roots. Animals have been observed upon such 'islands' many miles out to sea in recent times. G. G. Simpson has given this method of migration the felicitous name of the 'sweepstakes route', epitomising the essential ingredient of chance. Another sweepstake route crosses the Mozambique Channel from Africa to Madagascar; this route has been proposed as a means by which the earliest lemurs reached Madagascar to flourish there in isolation. However, evidence of continental drift makes it more probable that the lemurs of Madagascar evolved from a primitive placental primate stock which was occupying the region of Madagascar when it was still part of the mainland of Africa. The date of separation of Madagascar is put at 135 to 65 million years ago. The evidence for primate evolution would lead zoologists to put the date of final separation of Madagascar much nearer the 65 million mark than 135 million for there is no evidence to suggest that primates evolved much earlier than 70–80 million years ago..

Other land bridges shown in Figure 38

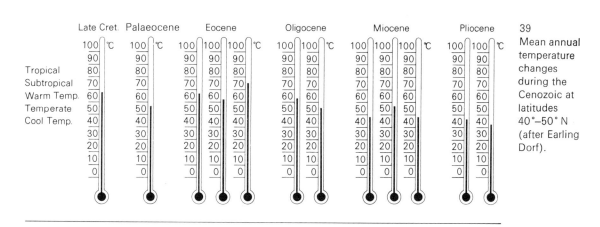

39 Mean annual temperature changes during the Cenozoic at latitudes 40°–50° N (after Earling Dorf).

include the Afro-Asian land bridge which connected Africa to northern India and the Far East during most of the Miocene epoch and which allowed an exchange of early Old World monkeys and apes between the two continents. There is reason to believe that early hominids such as *Ramapithecus* reached Asia from Africa via this route. A land connection between Africa and Europe in the region of Sicily or further west across the Straits of Gibraltar probably existed for short periods during the Oligocene and Miocene. Prior to these epochs, Europe and Africa were separated by the broad Tethys Sea of which the Mediterranean is now a tiny remnant. So much mountain building was going on in the bed of the Tethys that a brief connection could well have been established in the late Eocene via a string of islands. Migration of primates from Europe to Africa may therefore have occurred by a series of sweepstake routes, the animals hopping from island to island along the chain.

Land bridge connections between Southeast Asia and the islands of Sumatra, Java and Borneo were frequent occurrences as the sea rose and fell, especially during the Pleistocene. All these islands lie within the 100-fathom mark of the Sunda Shelf. Human and non-human primates must have used these bridges to colonise island after island ultimately reaching the Philippines along two narrow island chains stretching from north-east Borneo to Mindanao and Luzon.

Apart from the climatic and vegetational changes, there were other revolutions going on during the Cenozoic. The face of the land surface was also changing. The vast mountain ranges of the world today – the Rockies, the Andes, the Himalayas, the Alps, and the Atlas mountains – were products of the Cenozoic. This prolonged orgasm of mountain-building affected the shape of continents, the form of the oceans and the distribution of forests, grasslands, and deserts. During the early part of the Miocene epoch, the orogenic (mountain-building) activity of the Tertiary was in full swing. In Europe the Alpine peaks were rising out of the Tethys

Sea and, at the other end of this vast geological basin, the Himalayas were forming a formidable barrier between Asia and peninsular India. The rift-valley system of Africa, which extends 5000 miles from Tanzania to Israel and the Dead Sea, emitted its first rumbling warnings; and in the Americas, the Andes and the Rocky Mountain ranges were uplifted. New land surfaces were forming, and old land surfaces were being eroded and undergoing a revolution in the nature of their vegetational cover.

The influence of a cooling climate, which led to the withdrawal of tropical and subtropical forest belts towards the equator, plus the rain shadow effect which resulted in areas of relative dryness forming in the lee of mountain ranges, contributed to a widespread expansion of grasslands. Grasslands (called prairies or steppes in temperate latitudes, and savannas in the tropics) offered new evolutionary opportunities to mammals in general. The fossil record clearly demonstrates that horses, which until the Miocene were small, forest-dwelling, browsing mammals, were changing their habit, their habitus, and their habitat; they were rapidly becoming adapted to grasslands and to the grazing habit and were displaying adaptations of the teeth and limbs suited to their new environment and way of life (see Figure 7).

As we shall see, the effect of this climatic and mountain-building revolution was of considerable significance for the origins of the higher primates and man.

EMERGENCE OF THE PRIMATES

The primates were among the first mammals to put in an appearance. At this time, some seventy million years ago, primates-to-be were small, long-nosed ground-living animals rooting among the leaves of the forest floor for their insect food, distinguishable only by obscure characters of the teeth and skull from the other, contemporary, long-nosed, insectivorous creatures. With hindsight, some experts feel they can

40 Reconstruction from casts of fossil bones of *Smilodectes* (courtesy of Smithsonian Institution).

recognise these primates-to-be even though they possessed none of the arboreal characters by which we now recognise the order. They may well be right, but to those of us interested primarily in the adaptations of animals the Order Primates effectively came into being when the earliest representatives started to live in trees.

Plesiadapis of the Palaeocene is a most unprimate-like primate and is totally deficient in arboreal adaptations, while *Smilodectes* and *Notharctus* which appeared a few million years later were already advanced tree-climbers; clearly the acquisition of tree-climbing talents was a pretty rapid affair.

Arboreal characters can be briefly summarised:

1 Mobility of the hands and feet and particularly of the thumb and big toe which are well separated from the other digits and, in some primates, capable of being opposed.

2 Replacement of sharp claws by flattened nails, associated with the development of sensitive pads on the tips of the digits.

3 A shortening of the snout associated with a reduction in the apparatus and the function of smell.

4 Convergence of the eyes towards the front of the face associated with the development of stereoscopic vision.

5 A large brain relative to body size due to an increased complexity of the functional areas concerned with vision and touch.

6 An upright posture which is expressed in the ability of primates to sit, stand, swing or walk upright.

The lemur-like Eocene family the Adapidae (including the genera *Notharctus* and *Smilodectes*) possesses most of these arboreal adaptations: nails had replaced claws and sensory pads were developing on the finger tips, the eyes were converging and the snout was shortening, the brain was relatively large, and the locomotion pattern involved an upright body posture but with acutely bent hips and knees. This last feature alone merits particular interest because the upright posture is one of the hallmarks of mankind (see chapter 5).

Later forms such as *Necrolemur*, an early European tarsier-like primate, and *Hemiacodon*, a North American form, show similar postural patterns.

The next recognisable stage in the fossil record is seen during the Oligocene epoch. At present the relationship between the Oligocene and earlier Eocene primates is unknown. Most of our information about the Oligocene primates comes from a region of Egypt called the Fayum, now desert but once covered with dense tropical forest. Between 25–35 million years ago the Fayum was the home of an extraordinary variety of ape-like and monkey-like primates. Some, like *Parapithecus*, were probably destined to become true monkeys; some, like *Aeolopithecus*, to become 'half-apes' like gibbons; and some, like *Aegyptopithecus*, to become true apes like the chimpanzee and the gorilla. It has even been suggested very tentatively that one of these creatures called *Propliopithecus* represents the earliest known member of the human lineage. *Aegyptopithecus*, of which only teeth and jaws are known, has features strongly reminiscent of later apes, particularly the well-known Miocene fossil ape, *Proconsul africanus*.

During Miocene times volcanic activity, rift-valley formation and mountain-building – external expressions of continental movement – were in full swing. One consequence of this orogeny and the coincidental cooling of the earth's surface, which had been steadily proceeding throughout the Cenozoic period, was the spread of grasslands at the expense of forests. Grasslands (or savannas) offered new evolutionary opportunities to a variety of mammals, including the expanding population of primates in the rapidly shrinking forest zones. A few primate stocks, including the ancestors of man and the ancestors of the modern baboons, evidently reacted to the challenge of the changing environment. New horizons were opened up for our remote human ancestors and the way was paved for the evolution of mankind.

This chapter has been all about the environment and the changes that have taken place during the long journey of the primate order. All this may seem to you irrelevant to the main theme of this book which is concerned with 'monkeys without tails' – men and apes. But, believe me, it is not. You might as well try to understand the meaning of Christianity without ever having read the Bible as to comprehend the emergence of man without some awareness of the natural forces of the environment that brought it about.

Animals are the puppets of nature, they appear and disappear, they flourish and they wither away, they dance and posture or collapse into an untidy tangle of limbs at a twitch of a string.

4

The great divide

There have been plenty of 'great divides' in the history of the primate order or in any order for that matter. The partition of one to make two is the fundamental mechanism of life, it provides the basis for reproduction, growth and evolution. The fertilised ovum of a sexual species, whether it is fish, fowl or good red herring, divides into two daughter cells which subsequently each divide, divide and divide again until the organism has acquired its full complement of cells. Even when growth is complete, division of cells does not stop. Cell division is part and parcel of the maintenance and repair system of organisms whose components are subject to wear and tear. Chromosomes divide to build cells, cells divide to build organs and, ultimately, organisms, and organisms divide to produce species. A tree divides when it branches, roads divide, religious and political ideologies divide, and so do peanuts. The rate of division in some organisms is unbelievable. If bacteria were allowed to reproduce by division in a wholly uncontrolled fashion (which they are not for there are many natural mechanisms which serve to inhibit their growth) they would, within two days, achieve a mass equivalent to that of the earth.

But here we are concerned with the sort of division that takes place within an established group such as the primates. Reference to Figure 42 shows that evolutionary trees, like real trees, reflect natural processes only because they branch. A real tree without a branch is not a life-form but a telegraph pole. An evolutionary tree without a branch is – if one accepts the basic principle – an impossibility because evolution can only exist by virtue of variety. Almost by definition therefore branching or division is the essential accompaniment of the evolutionary process. Returning to

Figure 42 it can be seen that the major divisions are:

1 Division of basic primate stock into prosimians and anthropoids.

2 Division of anthropoids into New World and Old World groups.

3 Division of Old World groups into monkeys proper and the ape and human stocks combined.

4 Division of the composite ape and human stocks into two independent lines – apes and men.

The 'great divide' of this book, which is concerned with 'monkeys without tails', is the last one mentioned. From that moment on, the die was cast and the critical question of who would exhibit whom in whose zoo was finally resolved.

APES GREAT AND SMALL

In chapter 1 we took a walk round the 'primate zoo' but conspicuously our tour did not include the apes. We weren't ready for them then but now I think we are.

The apes are subdivided into two families, called in zoological language the Pongidae and the Hylobatidae. In common parlance the former are known collectively as great apes and the latter as lesser apes. The great apes are the gorilla, the chimpanzee and the orang-utan. The lesser apes are the gibbons and siamangs.

The chimpanzee and gorilla are more closely related to each other than either is to the orang-utan who is a kind of odd-man-out in this trio. We have chosen the chimpanzee for reasons already discussed but using different arguments we might equally well have chosen the gorilla. The differences between gorillas and chimpanzees are very superficial, so much so that one school of opinion places them in a single genus called *Pan*. I wouldn't go so far as this

41
An Orang-utan at Chester Zoo in bipedal posture (courtesy of D. Sorby).

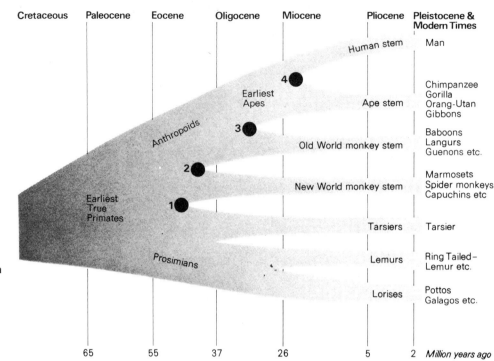

42
Evolutionary
tree of the
primates in a
simple form.
Large dots
indicate the
important
'divides'
(see text).

| Cretaceous | Paleocene | Eocene | Oligocene | Miocene | Pliocene | Pleistocene & Modern Times |

Human stem — Man

4 Earliest Apes — Ape stem — Chimpanzee / Gorilla / Orang-Utan / Gibbons

3 Anthropoids — Old World monkey stem — Baboons / Langurs / Guenons etc.

2 — New World monkey stem — Marmosets / Spider monkeys / Capuchins etc

1 Earliest True Primates — Tarsiers — Tarsier

Prosimians — Lemurs — Ring Tailed- Lemur etc.

Lorises — Pottos / Galagos etc.

65   55   37   26   5   2   *Million years ago*

however and I prefer to keep them in the separate genera – *Pan* (chimpanzees) and *Gorilla* (gorilla). Analogies between some other mammalian species might provide a perspective. Hares and rabbits belong to the same family Leporidae, hares belong to the genus *Lepus* and rabbits to *Oryctolagus*. On the other hand, lions and tigers of the family *Felidae* are placed in the same genus *Panthera*; and among the bears both the brown bear and the polar bear belong to the same genus, *Ursus*.

In spite of the superficial differences between them, chimpanzees have a greater claim to stardom than gorillas. We can identify more closely with the extroverted, playful chimpanzee than with the solemn, morose, and apparently menacing gorilla. Our attitudes in this respect are absurdly anthropomorphic and compounded of ignorance, myth and superstition.

The gorilla suffers from a fearsome reputation of rapaciousness, sexuality and brute strength. *King Kong*, the fictional monster who challenged the might of the United States Air Force from the top of the Empire State Building in the classic movie, epitomises for most of us the image of the ravening gorilla with strong sexual urges and saliva-flecked jaws. The media

have brainwashed us into believing that gorillas are brutish pseudo-humans of low mentality and vicious habits. Man of course is thoughtful, gentle and humane. If only this were true and not the other way around! Field studies of the gorilla in its natural habitat have exposed the myth, but have probably not totally obliterated it. Far from being a ravening monster the gorilla turns out to be an extremely peaceable creature, a vegetarian, an old softie whose bark is infinitely worse than his bite. The gorilla is his own worst enemy and like a lot of kind and gentle people he assumes in self-defence a curmudgeonly expression.

The popular image of chimpanzees is a bit of a fallacy too. Up to the age of four or five years chimps are fairly predictable and safe, but when they grow to maturity they tend, like humans, to put away childish things and turn to the more serious issues of life like sexuality and aggressiveness. Most of the chimps that appear in films, circuses, zoo tea-parties, and TV commercials are youngsters. Many people who see an adult chimpanzee for the first time in a zoo assume that they are looking at a gorilla, whom an adult chimpanzee quite closely resembles.

But before we concentrate on the gorilla and the chimpanzee, and particularly the

latter, it is important that we give some heed to the two apes that are not so closely related to us – the gibbon and the orang-utan.

GIBBONS AND ORANG-UTANS

Gibbons are slender, elegant animals with exceptionally long arms and hands and rather short legs. Their long arms reflect the special way in which they move through the trees. The gibbon's locomotor style has been given the name brachiation, or arm-swinging. These graceful creatures swing their way through the forests flowing from branch to branch with the beautiful rhythmical precision of the trained acrobat. But like all creatures that take risks of this sort 150 feet above the ground, accidents do happen; a dead branch, a slight error in the judgement of distance, and they come crashing to the ground. Fortunately the lower branches of the forest trees form 'safety nets', but nevertheless injuries in wild gibbon populations through falling are surprisingly high. One authority has estimated the accident rate as high as thirty-three per cent.

Gibbons and their close, but larger relatives, the siamangs, are inhabitants of the tropical rain forests and monsoon forests of South-east Asia. Gibbons are strictly arboreal animals only occasionally coming down to the ground. When they do, they move in a peculiarly human way. Having such long arms, which in fact touch the ground when they are standing upright, gibbons are obliged to walk vertically in a burlesque of the human walking gait, sometimes using the arms as a man employs a pair of crutches. The siamangs, which are considerably bigger than gibbons, differ from them in a number of relatively minor ways including the possession of a throat sac which when inflated by air during the vocalisation gives their calls a deep 'booming' quality.

Orang-utans, which once had a wide distribution on the continent of Asia, in southern China and Malaysia, are now only to be found in Borneo and in a small area in northern Sumatra. Orangs are about as large as gorillas, and the males may weigh as much as 400 pounds. As with gorillas, there is a great difference in size between males and females, the latter being about half the size of the former. In spite of their bulk, orangs are largely arboreal but not surprisingly show few of the 'flying trapeze' skills of the gibbon. Although they are classed as brachiators, orangs clamber about ponderously, using arms and legs indiscriminately for hanging, swinging and reaching. The feet are so hand-like that, effectively, orangs have four hands not just two. In fact nineteenth-century zoologists classified these animals as Quadrumana – or four-handed. Like all apes, orangs are vegetarians living on fruits, leaves, buds and flowers. Little is known about their social life but what we do know is rather curious (curious, that is, compared with the social behaviour of chimps, gorillas and man). The male orang-utan is the lord of all he surveys within quite large areas which form his sexual manor and all the females within it are his. His role is rather like a commercial traveller's who regularly visits his contacts, who are all separated from one another in different parts of the forest. Although this form of social organisation is familiar among lower primates, the mouse-lemurs and the galago for example, it is unique among the higher grades of the order. Orang-utans have very low population numbers in the wild; they are among the most endangered of all primate species. Possibly the social system of living orangs is adaptive to the peculiar ecological situation in which they find themselves today.

Just as the gorilla has been a focal point for many of the myths and legends of Africa so the orangs have filled the same role in Asia. The notorious Abominable Snowman, the *Yeti* of Nepal, Tibet and Bhutan, may well be based on the racial memory of the orang-utan, the traditional hairy man of far eastern woodlands. Although unknown today on the Asian mainland, there is fossil evidence that the orang survived there until perhaps two or three thousand years ago. It is even possible that the legend of the *Sasquatch*, the Bigfoot of British Columbia and north-western United States, is based on ancient tales of the giant man-like ape of the Szechwan province of China.

44 Phases of the brachiating swing of the gibbon (John Fleagle).

45 The gibbon in action in the wild.

GORILLAS AND CHIMPANZEES

The gorilla is the African counterpart of the orang-utan. One thinks of the gorilla and the orang as geographical analogues simply because they are both large animals. Size however is about all they have in common (apart, of course, from a common distant ancestry). The orang-utan is almost wholly tree-living, while the gorilla seldom gets off the ground except to sleep, and not always then.

The orang depends strictly on tree-products for its nourishment while the gorilla lives on ground plants. The orang lives a solitary existence while the gorilla is fairly gregarious and companionable; he is part of a troop of ten or more individuals of all ages and both sexes. Anatomically the coat of the orang-utan is red or reddish-brown while the gorilla is black heightened by swashes of grey in adult males; the orang has a very short thumb and an even shorter big toe while the gorilla, in keeping with his more terrestrial habit, has a longish thumb and a stout big toe; of all the great apes the gorilla approaches the human pattern most closely in this respect.

Like orang-utans, gorillas are an endangered species. Both the western 'lowland' gorilla of the Cameroun and the eastern 'highland' form of the Congo forest, Rwanda, Burundi, and south Uganda are at risk not so much from the depredations of man-the-hunter but from the most insidious destruction of forest habitat brought about by man-the-agriculturist and man-the-wood-user.

Although, taken overall, the anatomy and physiology of gorillas and chimpanzees is similar, they appear to be very different sorts of animals when judged on the basis of behaviour. Man in some aspects of his behaviour is very gorilla-like, in others he is essentially chimpanzee-like. The gorilla in man's nature is exemplified by the traditional image of the countryman – relaxed, easy-going in speech and action, given to brief, choleric outbursts but generally amiable and unaggressive and ready to meet life as it comes. The chimpanzee characteristics are more typical of the city dweller: dynamic, rumbustious and quixotic with a ready wit and a sharp eye for the main chance. As this analogy implies, temperamental and personality differences in chimps and gorillas owe much to their respective environments. This is true, but it is not the whole truth by any means as we shall see.

The theory that the mind of a newborn infant of any species is a *tabula rasa*, a clean slate on which the experiences of a lifetime are to be inscribed, sums up the attitude of John Locke (1632–1704), the English philosopher. Many social anthropologists hold tenaciously to Locke's theory today in spite of the contrary arguments of biologists. Few social anthropologists are prepared to admit that man has a cultural as well as a physical past but fortunately the happy few who have espoused this point of view are both literate and a highly outspoken lot. (See, for example, *The Imperial Animal* by Lionel Tiger and Robin Fox.) As a biologist, I see the slate that is the new-born infant's mind as being far from blank; rather it is like one of those 'magic' drawings which looks blank until a pencil is rubbed over its surface when it springs to life. The picture is there, but it needs the shading of experience to reveal it.

Individual gorillas and chimpanzees are born with all the quirks and oddities of their adult personalities lying fallow. The environment then moulds and modifies this latent source of uniqueness in a way which differs from individual to individual. In her classic observations of the behaviour of wild chimpanzees in the Gombe Stream National Park, Tanzania, Jane van Lawick-Goodall was struck, from the very beginning, with the extreme physical and personality variations of the group she was studying. This compelled her to employ human names to identify her animals – Flo, David, Flint, Goliath, Leakey and the rest. (See *In the Shadow of Man* by Jane van Lawick-Goodall, 1971.) It was natural to give human names to creatures whose very individual variations were reminiscent of the physical characters and behavioural idiosyncrasies of a group of human beings. Given a similarity of environmental pressures operating on a

46
A 'groomin
session'
amongst
adult males
(Baron.)

fairly homogenous group, like the Gombe Stream chimps, how can one doubt that the slates that were their minds at birth were already inscribed with a unique, genetic 'magic' picture that required only the touch of a crayon to bring it to life.

Thus it can confidently be argued that the personality differences of the gorilla and the chimpanzee are not simply related to the nature of their present environments. It is not enough to say that gorillas live in dense tropical forest (lowland and montane) and that chimpanzees dwell in more open forest and in woodlands and hence that in spite of a close genetic relationship their personalities are poles apart. Their personalities in my view are the products not only of their environmental present but of their environmental past.

As biologists we must learn to think historically. Henry Ford said 'history is bunk' and, in the context of building Model Ts, there is no reason to doubt that he was right – no precedent could possibly have helped him set up his production line. But with nature we are dealing with a totally different commodity. As George Gaylord Simpson has said, 'The past of animals is one of the determinants of their future.'

One scientist, the Dutch zoologist Adriaan Kortlandt, has for many years been thinking 'historically' about chimpanzees. Why, he asks, are chimpanzees so different from gorillas, why are they, superficially, so man-like? Why, when faced with predators like leopards, do they behave in an aggressive man-like fashion, when gorillas in similar circumstances would adopt a purely defensive role? Furthermore, why do chimpanzees resort to the use of weapons when faced by such predators? Why are they prone to use naturally occurring objects as tools? Why do they resort so often to two-footed walking? Why are they carnivorous (when the opportunity offers) while gorillas remain dedicated vegetarians? These are all good questions, possible answers to which have suggested to Kortlandt that chimpanzees have not always been the forest-dwellers that they are today but that in the past they were living in more open country, in the savannas and grasslands of Africa. Kortlandt believes that with the coming of man with his clubs and spears, chimpanzees were forced back into the forests where they are still to be found. Many aspects of their present behaviour, Kortlandt avers, are attributable to those far-off days when chimpanzees were the inheritors and men were the lesser primates. He cites the use of sticks against predators, the upright posture, and the occasionally expressed carnivorous propensities of chimpanzees as supporting evidence for their ancient role as the up-and-coming 'humans'. Thus, he continues, with the coming of 'true' man, the chimpanzees became dehumanised. The fallacy here is that it is only possible to dehumanise a creature that is already human. Super-chimps they may have been, but they were apes not men. So if there is any truth in Kortlandt's hypothesis, the process of degradation was one, not of 'dehumanisation', but of 'desuperchimpanisation'.

I think it likely that Adriaan Kortlandt is correct in stressing that the chimpanzee today is a very different animal from the chimpanzee of yesterday, but I think he is wrong in his insistence that chimpanzees ever occupied a human-like niche in nature. Men, in the sense of the members of the family Hominidae, have been in existence for as long as apes; the man-like niche has been occupied for as long as the ape-like niche. The fallacy that man has descended from an ape has been discussed elsewhere and shown to be entirely erroneous. When the ape and the human lines separated the common ancestor of the two stocks was neither an ape nor a man. If pressed to describe the appearance of such a common ancestor I would be compelled to supply a series of features that would leave you in no doubt that the common ancestor was more like a monkey than anything else.

EVOLUTION OF THE APES

After a brief respite from the evolutionary history of the primates we pick up the story where we left it in chapter 3. We find ourselves at the beginning of the

47 Reconstruction of *Proconsul africanus* based on known characteristics, drawn by Maurice Wilson.

geological epoch known as the Miocene which began about twenty-three million years ago and ended eighteen million years later.

The Miocene was the heyday of the superfamily Hominoidea which comprises the apes and man. The Miocene was the epoch when it was all happening. During this time the geography and climate of the globe was undergoing a tremendous revolution and if you have followed the principal argument of this book you will appreciate that environmental changes are inevitably followed by physical and behavioural changes in animals. The Miocene was a period of volcanism and mountain-building that more or less defined the surface texture of the world as we know it today. The vegetational changes that followed the geological revolution known as the Cascadian are fundamentally important. Climatically the world was cooling and rainfall was becoming more seasonal as the influence of ocean currents and prevailing winds affected precipitation patterns in the middle latitudes. Only in the tropical zone, which forms a girdle around the belly of the earth with a width no greater than 10° of latitude, were conditions reasonably static.

The result of these changes on vegetation patterns was that great tracts of forests disappeared, partly as a result of dryness and partly as a consequence of the rain-shadow effect of newly formed mountain ranges. In nature, as in human affairs, there is always a replacement handy, an understudy waiting in the wings. 'The King is dead, long live the King' is a principle that permeates all life systems. Grasses are monocotyledons of the sub-order of angiosperms, the flowering plants, and have been in existence since the Jurassic, but it was not until the late Oligocene and the early to middle Miocene that they really came into their own. Grasslands are known regionally by such terms as savannas, llanos, prairies, meadows, etc. Differences between them are due principally to local rainfall conditions. Tropical savanna for instance is subject to annual burning in many parts of Africa during the dry season. The origin of grasslands is a subject of considerable controversy

and nobody is quite certain as to how much they owe their existence to nature, and how much to human agricultural methods such as annual burning. However there is universal agreement that some twenty-five million years ago the grassland : forest ratio was increasing during the Oligocene-Miocene epochs.

This was the period when a new sort of primate was evolving – a savanna primate. Today, ground-living species are – in terms of population numbers and intelligence – the most successful of all primate species. They include the wily macaques, the regimented, orderly baboons, and the unpredictable species called man. Fossil apes are known in large numbers from the early Miocene fossil sites in Kenya and Uganda, particularly around the shores and on some of the islands of Lake Victoria. The genus most commonly represented is known as *Proconsul* and includes three species: *P. africanus*, *P. nyanzae* and *P. major*. The ecology of the areas in which their fossil remains are found suggests a tropical mosaic of forest-grassland well watered by streams and heavily fringed by gallery forest. Some areas were on the slopes of active volcanoes heavily forested in the manner of the montane habitat of the eastern ('highland' or 'mountain') gorilla of today; it is from these latter sites that the remains of the largest *Proconsul* species, *P. major*, are found. The anatomical characters of *P. major* combined with inferences of habitat strongly suggest that a gorilla-like form was already in existence in the early Miocene. If it seems unlikely to you (as it does to me) that the genus *Gorilla* was already in being twenty million years ago one should remind oneself that given environments tend to promote the evolution of characteristic life-forms. A gorilla is not so much a species, it is more a way of life. It is probable that a gorilla-type of higher primate has cropped up more than once in the fossil record of the living apes.

Remains of the smallest of the three *Proconsul* species, *P. africanus*, are by far the commonest primates in the Miocene deposits of East Africa. A complete skull was recovered in 1948, and in 1951 parts of a

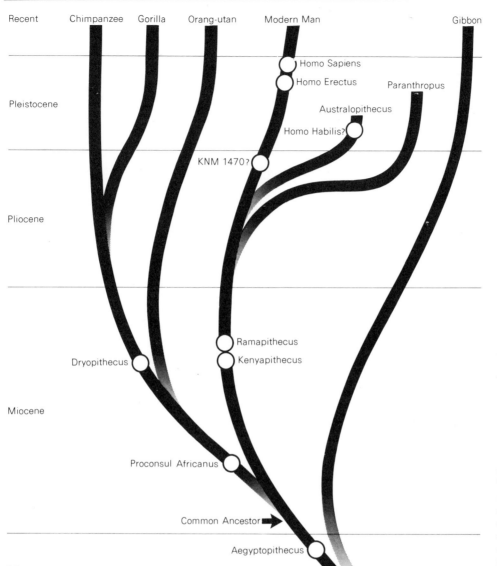

Recent — Chimpanzee — Gorilla — Orang-utan — Modern Man — Gibbon

Pleistocene

Homo Sapiens
Homo Erectus
Paranthropus
Australopithecus
Homo Habilis?
KNM 1470?

Pliocene

Ramapithecus
Kenyapithecus
Dryopithecus

Miocene

Proconsul Africanus

Common Ancestor →

Aegyptopithecus

Oligocene

48
A simplified
evolutionary
tree of the
apes and
man. Like
all such
diagrams,
its basis
is highly
speculative.
We do not
really know
enough to
draw such
solid,
continuous
lines.

skull and an almost complete upper limb including the hand came to light on Rusinga Island on the Kenya side of the lake. These specimens combined with several hundred other fragments of jaws, teeth and limbs have provided the basis for two major monographs on this species published by the British Museum (Natural History). *Proconsul africanus* was a small creature weighing between 23–35 pounds. It was quadrupedal in gait and about as agile and acrobatic in trees as a living spider monkey, which is saying a lot. Although it was monkey-like in its bodily form it is decidedly ape-like in the pattern of its teeth; in fact the teeth of the lower jaw

are very similar to those of a living chimpanzee. In the upper jaw the molars show certain minor specialisations which may indicate that *P. africanus* was slightly 'off' the mainline of chimpanzee evolution.

The present view is that the *Proconsul* group were descendants of the Oligocene primate from the Egyptian Fayum (see page 72) called *Aegyptopithecus*. As far as the relationships between *Proconsul* and later apes are concerned we lack the necessary fossil confirmation to be dogmatic, but it is most likely that they represent the ancestral stock of modern African apes. *Proconsul africanus* has also been proposed as the common ancestor of man and apes. It is possible, of course, but unlikely.

In 1932 a graduate student of Yale University, G. E. Lewis, discovered part of an upper jaw of an ape-like creature in the Siwalik Hills, India. He called it *Ramapithecus* and expressed the view that *Ramapithecus* was an early member of the human family, but the scientific climate of the period was not propitious for such claims and the specimen faded into obscurity for the next thirty years. The specimen was resurrected in 1961 appropriately enough by another Yale scientist, Dr E. L. Simons, who reinstated *Ramapithecus* as an ancestral human. Other specimens have now been recognised in existing fossil collections that are attributable to the same genus. *Ramapithecus* has also been found in Africa where it was given the name *Kenyapithecus* by its discoverer, Dr L. S. B. Leakey. In spite of the name, *Kenyapithecus* is clearly identical with *Ramapithecus* which, being the older name, has priority. *Ramapithecus* of India and Africa represents the earliest evidence of the human line in the fossil record and dates back some ten to fourteen million years. The 'human' characters of *Ramapithecus* relate to the jaws and teeth only as no postcranial remains of this creature have yet been identified. The human status of *Ramapithecus* is being strongly challenged at the moment (see page 111).

Other fossils that are undoubtedly ape-like in the widest sense of the word include *Dryopithecus sivalensis* from the Siwalik Hills late-Miocene deposits, a possible ancestor for the orang-utan; *Gigantopithecus bilaspurensis* from the same region and from areas in China, a giant-jawed vegetarian creature that is neither ape nor man but something in between; *Pliopithecus*, a gibbon-like fossil from the Miocene of France, Hungary and Czechoslovakia; and, finally, *Oreopithecus*, the most enigmatic fossil of all. *Oreopithecus* has been known for about 100 years but the bits and pieces then known did not tell us very much except that it might be some kind of an ape. In 1958 a complete, although grossly compressed and distorted specimen was discovered lying *in situ* in the roof of a tunnel in a brown-coal mine at Baccinello in Tuscany. The specimen, flattened to the thickness of a piece of cardboard, was carefully removed in a single block and transported to Basle, Switzerland, where it has been extensively studied by Dr Johannes Hurzeler. Hurzeler holds the opinion that *Oreopithecus* belongs properly to the Hominidae, the family of man, on the grounds that its skeleton shows many adaptations to bipedal walking, but he has few supporters. The critics point out that the teeth are quite unlike those of pongids and hominids. In spite of certain bodily adaptations which might suggest that *Oreopithecus* was capable of swinging by its arms in the trees and walking in a two-footed way upon the ground it clearly is neither a hominid nor a pongid. As in the case of *Proconsul major* (see above) the environment can play a very devious role in shaping an animal.

The 'great divide', the separation of the ape and human stocks, was a critical event in the history of the primates but unfortunately we shall never know exactly when it took place. The reasons for this are threefold: first, evolutionary change is not an instantaneous event like a chemical reaction but a slow and gradual process that spreads over many generations; second, the fossil record at the critical period of time is so poor that there is little material evidence to guide us; and, third, the methods available for dating such material as we do have are rather inaccurate, the margin of error being of the order of one or two million years for dates of ten million years and over. Where more recent events are concerned, say within the last 10,000 years, the margin of error using radioactive dating techniques can be narrowed very considerably; here the margin of error is a matter of 200–300 years.

As a result of all this maddening poverty of information we are driven to supplying broad estimates of the timing of the 'great divide', conservatively bracketing the targets by generous upper and lower limits. I say 'targets' because there is no universal agreement about even the epoch during which this event took place. There are at least three current schools of thought. They can be referred to as the late-late school, the late

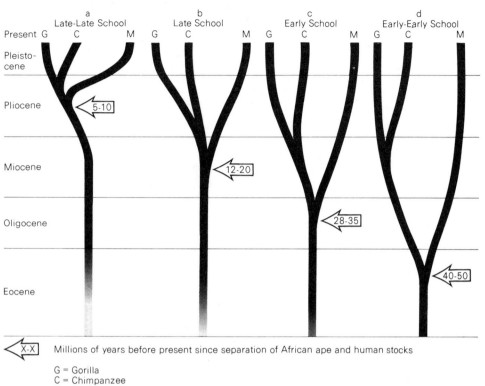

Present G   C   M   G   C   M   G   C   M   G   C   M

Pleisto-
cene

Pliocene

Miocene

Oligocene

Eocene

5-10

12-20

28-35

40-50

X-X Millions of years before present since separation of African ape and human stocks

G = Gorilla
C = Chimpanzee
M = Modern man

49
Diagram summarising the principal views of the time of separation of the human and the African ape stocks.

school and the early school. A fourth school, the early-early, enjoyed a brief period of popularity but is probably only of academic interest today (Figure 49).

The early school holds that the separation took place in the Oligocene epoch between twenty-eight and thirty-five million years ago. There is some, as yet rather insubstantial, evidence for this. *Propliopithecus* and *Aegyptopithecus* of the Fayum Fauna of Egypt (see page 72) show the sort of differences in their teeth that could be interpreted as anticipating the dental differences between men and apes today. Unfortunately only the teeth and jaws are available so there is no means of determining whether the relevant locomotor differences were beginning to show themselves at this level.

The late school favours a Miocene date some twelve to twenty million years ago. A possible candidate for common ancestry would be *Proconsul africanus* of the early-middle Miocene. For reasons already discussed, *P. africanus* would seem already too specialised along the ape-line to fill the role of a common ancestor. More generally

acceptable is the idea that the common ancestor is to be found among the successful genus of Miocene apes known as *Dryopithecus*. During the late Miocene and early Pliocene the genus was widespread over Europe, Africa and Asia. The presently held view is that from the 'dryopithecine complex' three major types arose:

1 Ape-like forms ancestral to modern gorillas, chimpanzees and orang-utans.

2 Intermediate forms of heavily built vegetarian creatures, somewhat ape-like yet somewhat human-like, which ultimately became extinct (*Gigantopithecus*).

3 Man-like forms (e.g. *Ramapithecus*) which evolved into the early species of near-man such as the Australopithecines of South and East Africa.

The late-late school is the brainchild of certain anthropologists involved in the study of the comparative biochemistry of primates. Based on the study of blood albumins of living primate species, V. M. Sarich and A. C. Wilson of the University of California at Berkeley have drawn certain conclusions about the time of divergence of various fossil primate species. In order for their arguments to hold water they have had to assume that albumins in the blood have undergone evolutionary change at a constant rate; it is this assumption that is under attack. On the face of it, such 'regularity' appears to conflict wholly with the principle of random mutation which is the axis and pivot of the Darwinian theory of evolution to which most zoologists subscribe. On such grounds as these biochemical anthropologists place the great divide (between the ancestors of the African apes on the one hand and the ancestors of man on the other) at approximately five million years ago. As must be apparent, such a date conflicts with the fossil record as we know it, given even the limitations of dating methods discussed above.

Whichever turns out to be the 'right school', and it may be a long time – if ever – before we can be certain, it is clear that the Miocene-Pliocene epochs were great times for the apes. From the enormously wide geographical distribution (from Spain, through France and Germany to Russia, and from Africa through the Middle East to India and China) it is evident that the Dryopithecines were a highly successful group. It has been calculated that before a recent reappraisal of this group, which resulted in the whole assemblage being placed in three genera (*Dryopithecus*, *Ramapithecus* and *Gigantopithecus*), fifty species distributed among twenty-five genera had been described from Spain through Germany, to Russia, Asia and points east.

This being so, whatever happened to the successful apes to turn them into endangered species today? What happened is what happens to ham in a ham sandwich; it becomes compressed to wafer-like dimensions squeezed between two hulking slabs of bread. Man in this instance was the bread, or at least one of the slices; the other slice was the changing environment of post-Miocene times. The reduction of forests and the coming of man, between them, put these rustic creatures out of business. Rather late in the day we are now doing our best to make amends for the sins of our ancestors by applying conservation measures to those creatures that our progenitors unwittingly steered into evolutionary jeopardy.

HOW TO TELL YOUR FRIENDS FROM THE APES

If you have ordinary social aspirations and do not like to offend your acquaintances unduly then some guidance on this delicate subject could be useful to you. The differences between your friends and the apes are either negligible or profound depending on your attitude of mind. Viewed dispassionately, chimpanzees and man share a vast number of anatomical and physiological characters but behaviourally and culturally they are poles apart. Viewed passionately there are many aspects of the behaviour of apes which are highly evocative of similar patterns of human beings, though in a modified form. As usual the truth lies somewhere in between these two extremes. It is dangerous to draw behavioural conclusions from the study of human behaviour and attribute them entirely to the inheritance of certain genetically established patterns of behaviour. Equally, it is intellectually indefensible to assume that human behaviour has not been affected by the past history of man's non-human ancestors. We are back, essentially, to the nature-nurture type of argument that has already cropped up on a number of occasions in this book. However, in the following section we are concerned only with observable and measurable differences and similarities between apes and man.

Termite-'fishing'. A slender stem of appropriate length and thickness is selected and trimmed before use. This behaviour constitutes a primitive type of tool-making.

Chimpanzees of Gombe Stream
National Park and Jan Van Lawick-
Goodall

Above: Adult male Bornean orang. *Below:* Female in characteristic hanging posture, carrying young infant.

DISTRIBUTION, ECOLOGY AND DIET

CHIMPANZEE

Distribution: East, Central and West Africa. Found in most forested and woodland areas between latitudes of 10°N and 7°S.

Habitat: Tropical forest and woodland savanna.

Diet: Essentially vegetarian including fruits, leaves, flowers, buds, bark, ground plants, insects. Eat meat occasionally including new-born antelopes, monkeys, bush-pig, etc.

Rhythm of Activity: Diurnal. Chimpanzees wake at sunrise and retire at sunset. Afternoon siestas are routine.

MODERN MAN

Distribution: Global. Found in most habitable areas. Occasionally to be found visiting regions that cannot support life without artificial aids, e.g. high mountains and polar ice-caps.

Habitat: Tropical forest, woodland savanna, open savanna, temperate grassland and forest, boreal forest, tundra, mountains and deserts.

Diet: Omnivorous. Man is a gatherer of fruit and vegetable products, a hunter of flesh, a breeder of edible livestock, and a fisherman. Man also consumes quantities of synthetic foods.

Rhythm of Activity: Diurnal. Primitive man lives by the sun, but in northern latitudes lives by the clock. He sleeps on average 8 hours per 24-hour day, and in tropical latitudes he also favours an afternoon siesta.

VITAL STATISTICS

Weight: Average weight—
   of new-born 1.58 kg (3 lb 8 oz)
   of adult male 54 kg (120 lb)
   of adult female 47.7 kg (105 lb)
Height (standing upright): Adult male 3ft 6 in (105 cm)
Gestation period: 238 days
Female puberty: 7–8 years
Life span: 40 years

Weight: Average weight—
   of new-born 3.29 kg (7 lb 3 oz)
   of adult male 65 kg (143 lb)
   of adult female 58 kg (128 lb)
Height: Adult male 5 ft 8 in (171 cm)
   Adult female 5 ft 4 ins (159 cm)
Gestation period: 266 days
Female puberty: 14–15 years
Life span: 65–75 years

ANATOMY

Anatomical differences are principally of degree (quantitative) rather than of kind (qualitative) as might be expected in two closely related species such as the chimpanzee and man. Historically, chimpanzees are arboreal animals, while man is a ground-liver, and the main anatomical differences of degree listed below reflect this environmental dissimilarity. The locomotion of apes is forelimb-dependent while that of man is hindlimb-dependent. Seen in this light, the distinctions listed below are not difficult to understand.

External characters: Hair is retained over the whole body except the palms of the hands and the soles of the feet. It is long and straight varying in colour from grey to black to dark brown; a reddish tinge is sometimes seen. The beard is not pronounced and baldness of the front of the head is common in males and females from maturity onwards.

External characters: Hair is restricted to certain regions: the head, the armpits and the groin although in males the chest and the back are sometimes almost as hairy as in chimpanzees. Over the rest of the body there are fine, relatively colourless hairs, which are usually thicker and darker in males than females. Man cannot be said to be 'naked', but he certainly does not possess anything that one could call a pelt. Baldness is more marked in males than females (see Figure 54).

Adult male gorilla, western variety.

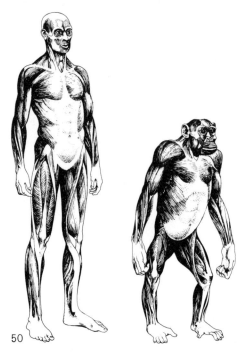
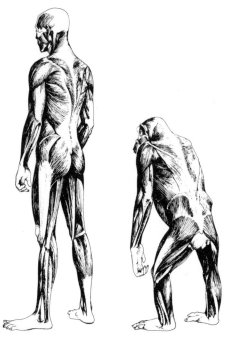

50

CHIMPANZEE

Limb proportions: Long bodies, rather short legs and long arms; arms are ten per cent longer than legs.

Posture and gait: Normal stance on the ground is a modified quadrupedal posture in which (by virtue of long arms and short legs) the head end is raised higher than the rear end. Capable of standing and walking upright for short distances on two legs but posture is rather stooped and the walk ungainly involving a large element of side-to-side rocking and weaving. When sitting, the attitude is very human-like. When moving in trees a variety of postures are adopted including an arm-swinging progression called *brachiation*.

Hands and feet: Hands are long in terms of primate standards, particularly the fingers, but the thumbs are short. The percentage length of the thumb of the index finger ('opposability index') is 42. Manipulative ability is only moderate; the principal functions of the hands are twofold: to provide a strong power grip for arboreal locomotor activities; and to provide support for the body when moving quadrupedally on the ground, the so-called knuckle-walking posture.

Hands of chimpanzees are well-endowed with sensory endings in skin and deeper tissues but tactile functions are neither as precise or discriminating as in man; the pain

MAN

Limb proportions: Long bodies, very long legs and relatively short arms; arms are thirty per cent shorter than legs.

Posture and gait: Normal stance is upright, a posture in which the weight of the super-incumbent body is finely balanced over the feet. To achieve this the vertebral column is compounded of a series of curves so arranged that a plumb-line dropped from the base of the skull passes through the feet just in front of the ankle joints. When man walks he progresses smoothly without undue side-to-side rocking or weaving due to a series of anatomical modifications of the pelvis and lower limb designed to render human walking as economical (in terms of energy consumption) as possible.

Hands and feet: Hands are of average proportions in primate terms. The particular feature of human hands is the relative length of the thumb. 'Opposability index' is 65 and, thus, manipulative ability is highly developed. Man has two major grip patterns, the power grip and the precision grip (see chapter 5). His hands are not normally used for locomotion except in infancy and in exceptional circumstances in adult life. An equally important function of the human hand is to provide sensory contact with the environment through nerve endings in the skin and deeper tissues sensitive to external stimuli of touch, temperature and pressure. Excessive stimula-

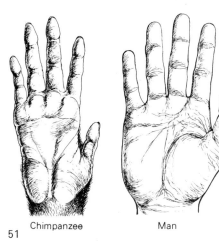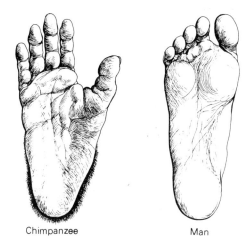

51 Chimpanzee   Man   Chimpanzee   Man

CHIMPANZEE

threshold is probably similar however.

Feet are prehensile, that is to say the big toe is well separated from the remainder and is highly mobile and can be used for grasping. The foot is rather 'flat' in human terms, lacking a longitudinal arch.

Jaws and teeth: 32 teeth in adult dentition as follows: 2 incisors, 1 canine, 2 premolars, 3 molars, in each half of each jaw. The dental arcade is somewhat rectangular. Canines project above and below tooth row and are stout and pointed, more so in males than females. Molars increase in size from front to back and carry cusps that are high and pointed.

Jaws are strong, rather long and protrusive; a chin is totally absent.

MAN

tion in all of these categories produces pain.

Human feet are essentially walking organs. Their grasping faculties are reduced to a minimum because the big toe lacks the essential feature of mobility; it is aligned alongside the other toes and acts with them in the performance of human walking and striding.

Jaws and teeth: Number and distribution of teeth as in chimpanzee. Teeth are arranged in a parabolic curve. Human canines do not project and are similar to incisors in shape and function. Sexes do not show size differences in canine size. Molar teeth decrease in size from front to back and cusps are rather low and rounded. Third molars (wisdom teeth) often absent.

Jaws are less powerful than in chimpanzees and there is a well-pronounced bony chin.

52

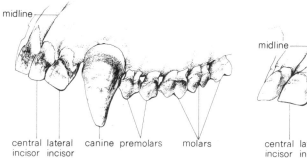

midline

central lateral canine premolars molars
incisor incisor

midline

central lateral canine premolars molars
incisor incisor

CHIMPANZEE

Brain and braincase: Brain is small although bigger relative to body size than in any monkey. Size of brain expressed in volumetric terms averages 394 cubic centimetres (cc). The braincase is low and flattened particularly in the frontal region where a forehead is absent.

Above the eyes is a continuous and strong bony crest. The opening for the spinal cord (foramen magnum) on the base of the braincase is directed downwards and backwards.

MAN

Brain and braincase: Brain is both absolutely and relatively large. The average volume for modern man being 1350cc, over three times as large as the chimpanzee.

The braincase forms a high vault, the frontal region being vertical forming a 'forehead'. There is no bony crest above the eyes. The foramen magnum is directed directly downwards – an adaptation for the upright posture.

53

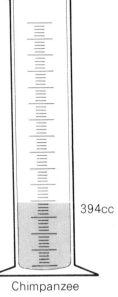

394cc

Chimpanzee

1305cc

Man

Vertebral column: Although there is considerable variation, chimpanzees have the same number of vertebrae in the spinal column as man but they differ numerically from region to region. There are usually less in the lumbar region (4) and more in the sacral (6). The lumbar region is either flat or is curved with its convexity backwards.

Vertebral column: In man both the lumbar and sacral regions contain 5 vertebrae. The lumbar region is always curved with its convexity directed forwards. Again, this is an adaptation for the upright posture which is missing in chimpanzees.

PHYSIOLOGY

Pulse rate: 70–80 per minute
Respiratory rate: 18–20 per minute
Body temperature: (**rectal**) 98.6°F
Blood pressure: 140–160/80 mm Hg
Blood groups: A and O
Chromosome number: 48
Menstrual cycle: 35 days

Pulse rate: 70–80 per minute
Respiratory rate: 16–20 per minute
Body temperature: (**oral**) 98.4°F
Blood pressure: 120/80 mm Hg
Blood groups: A, AB, O and B
Chromosome number: 46
Menstrual cycle: 28 days

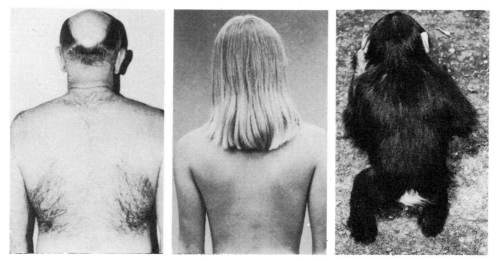

54 Anatomically Man is not hairless. Both males and females have hairs all over their bodies but apart from the head, armpit and groin, they are generally short, fine and colourless. They can have little function in protecting the skin against heat loss and injury. The thickness, length and visibility of the body hair varies both between the sexes and within them. The chimpanzee has longer, darker, thicker hair than Man, but, numerically in terms of hairs per unit area, there is little difference.

A comparison of the anatomical and physiological characters of chimpanzees and man reveals the very close similarity of the two species in every department. This is particularly true of the physiological 'norms' such as the blood pressure, respiratory rate, pulse rate and menstrual cycle, etc., which are subject in both species to considerable variation between individuals, even in normal health.

Chimpanzees are subject to most of the diseases that man is heir to. They develop bacterial infections like dysentery and puerperal (or 'childbirth') fever, virus infections like chicken pox, measles and mumps, and even certain congenital conditions such as mongolism. A few years ago the chimpanzees in the Gombe National Park in Tanzania suffered a disastrous epidemic of poliomyelitis, a disease which they had caught from Africans of a local village who were suffering a severe attack. From various accounts of the course of the disease it appeared to follow a typically human pattern. It seems that by definition a good man-doctor is a good chimpanzee-doctor as well.

BEHAVIOUR OF MEN AND
CHIMPANZEES

It is easy enough to tabulate data for comparison when the elements involved vary in a quantitative way such as height, weight, gestation period and longevity. It is not so simple when one is dealing with matters that differ qualitatively between two species. For example, how can the apparently promiscuous mating behaviour of chimpanzees be compared with the pair-bond mating which is normal for man? Because mating behaviour in man is so powerfully moulded and modified by his culture, the form it takes differs profoundly from similar behaviour in a non-cultural animal like the chimpanzee. Sex in chimpanzees is a simple exercise in procreation (which is biological), while in man it is often an expression of love (which is cultural) having nothing to do with procreation at all. The same difficulty of attempting to compare behavioural patterns that are not biologically comparable is encountered in many aspects of social behaviour.

As behaviour is an adaptive response of an animal to its environment, it is not surprising that, bearing in mind the profound differences in life-style of their immediate ancestors, the behavioural patterns of living chimpanzees and man are so unalike in many respects. It is perhaps more surprising that man and chimpanzees share as many behavioural characteristics as they do, but it mustn't be forgotten that ecologically dissimilar as their recent pasts have been, there was a time in the more remote past when it is believed they shared a common way of life and thereby acquired homologous behavioural traits.

Considering how closely related the chimpanzees (and gorillas for that matter) are to us it is disappointing to say the least that we cannot learn more about ourselves from investigating their natural lives. It would be nice to be able to say: 'Ah! Now I understand why man behaves as he does.' As a result of anatomical and physiological comparability of the two species, we can do all sorts of things like putting chimpanzees into orbit, testing organ transplant techniques and investigating the effects of new drugs, thereby obtaining invaluable information, but there is disappointingly little that we can learn about social behaviour.

Quite apart from behavioural differences that stem from dissimilar ecologies, there is a further factor that makes behavioural projections unreliable. Man's social behaviour is so obviously culturally determined that many social anthropologists hold the view that biological factors have no part to play at all. The philosopher Ortega y Gasset epitomised this view when he asserted: 'Man has no nature, what he has is history.' Geneticists on the other hand, fully aware of the role of the genes, have wholly abandoned the nature-nurture (the 'either-or') concept and concern themselves with assessing the relative importance and the interactions of the two forces in the composition of what we call human nature. Anatomists, zoologists, and physical anthropologists have also abandoned any idea they might have had of the uniqueness of human behaviour; for them structure and behaviour are part of one indissoluble whole and to admit to uniqueness of behaviour would be to deny the heritability of structure, which is of course absurd.

Social behaviour in man is partly a biological and partly a cultural phenomenon. Man may be unique in possessing culture but he is at one with all higher primates in possessing a social system. In fact one can go further and state that – barring language and language-dependent behaviour – there are no basic patterns of human behaviour that are not found in one or other of the Old World monkeys or apes. Human societies are designed on biological principles however much they have been camouflaged by the 'buttons and bows' of cultural couturiers.

Some Old World monkeys, like the Japanese macaques, and apes, like the chimpanzee, are believed to avoid mating with their mothers; and this is in spite of never having heard of Oedipus or his complex or the moral principle which in our society forbids the practice of incest. There is some evidence amongst chimpanzees that even exogamy is a natural practice. In human societies exogamy, or marrying-out, is a basic principle – the corrollary of the incest taboo. Like the incest taboo, this simple biological process of nonhuman primates has become complicated by a whole series of conventions and ritualistic procedures – the 'buttons and bows' in fact.

It is apparent that what one might call the zoological approach to the evolution of human behaviour is fraught with difficulties because of the qualitative difference between 'animal' and 'human' behaviour, but as we shall see it is in fact the most promising approach of all.

It can now be appreciated why a neat, tabulated comparison of chimp and human social behaviour is not a feasible proposition, except in certain rather limited ways that concern those behavioural characters acquired when man and chimpanzees were part and parcel of the same hominoid stock in the distant past. Such characters of common inheritance as they are called concern feeding behaviours, maternal care, facial expression and other non-verbal forms of communication.

We are in the midst of a naturalistic revolution. Conservation, macrobiotics, ecology, pollution are fashionable words much in the thoughts of many people, particularly the younger generation who, rightly, are the most concerned. Public interest generates public funds and consequently scientific research has benefited and our knowledge of human prehistory is accumulating at an unprecedented rate. Natural history, once the pursuit of clergymen and the eccentric gentry, fell into disrepute in the latter half of the nineteenth century. Now, once again, it has become respectable but there has been a change; its disciples are no longer amateurs, but professionals. It seems appropriate, therefore, to look at some of the ways in which the natural history of man is being investigated.

Mankind today is represented by a single species, *Homo sapiens sapiens*, to give him his full title. Unlike the non-human primates which include in their ranks primitive as well as advanced forms there are no truly primitive forms of man in existence. As each new and improved 'model' came on the market the old versions, so to speak, became extinct.

What about the so-called primitive races of mankind like the Bushmen, the Hadzas, the Australian aborigines, the 'lost' races of the Philippines, New Guinea and the Amazon basin? Surely these 'stone-age' people surviving on a hunter-gatherer economy can provide some of the answers to the behaviour of our lineal ancestors? Sad to say, they cannot. Just because a tribe of Indians lives deep in unexplored territory it does not mean that evolution has passed them by. In their own peculiar way they are as specialised as we are and far removed from the model we are seeking – a truly primitive form of man. This is not to say that the field study of Bushmen and other primitive tribes does not bring its own rewards in terms of the better understanding of the ecological adaptability of man and the sort of diet that will keep him alive and healthy under adverse natural conditions.

Failing the living primitives, the only places that we are likely to be able to find primitive forms of man are on fossil sites. Unfortunately not only are human fossils rather sparse but the potential for drawing behavioural conclusions from a mixed bag of petrified bones is severely limited. It is possible to make certain deductions of a simple sort from the debris associated with the living sites of fossil man. This sort of evidence reveals the habits of the occupants in much the same way as the detritus of a human picnic area provides insights into the idiosyncrasies of holiday-makers. Fossil sites can tell us something about diet, about the use of fire, about toolmaking, and about hunting behaviour, but they reveal very little about the form of human society and nothing at all about social motivations. Furthermore, living sites are not typical of man before about three million years ago; the earliest hominids are presumed to have been nomadic groups that roamed the savannas in pursuit of rich feeding grounds and water-holes. Even at Olduvai Gorge where living sites are known, they were probably little more than temporary camps, although at one site there is evidence of a crude form of semicircular shelter having been constructed. It is unlikely that earlier hominid bands in the pre-toolmaking, pre-hunting stage of evolution would have left any permanent marks of their passage.

Chimpanzees are obviously the logical candidates for a zoological approach to the problem of the evolutionary history of human behaviour, but their potential is limited by the ecological differentials that have been operating on the two stocks since the 'great divide'.

Man is a ground-living primate, as his bipedalism attests, so perhaps it is among other ground-living Old World primates that we should look for inspiration; inevitably we are led to the most interesting group of primates living today – the baboons and the macaques.

The savanna is well stocked with primates in terms of numbers rather than variety. In Africa there are the baboons, a blanket term which has been used to include at least three genera which are more or less closely related. These are the common baboons (*Papio*), the drills and mandrills (*Mandrillus*), and the geladas of central Ethiopia (*Theropithecus*). Although baboons live in open savannas, mandrills in tropical rain forest, and geladas on mountain moorlands, they share a broad ecological category, they all live on the ground. In fact geladas, in whose habitat trees are absent, must be the most ground-bound of all primates including man. These 'baboons' differ only in minor and rather superficial anatomical details from each other. Behaviourally they are more variable as might be expected from the disparate nature of their environments, but nevertheless they are more behaviourally similar to each other than any one of them is to their arboreal relatives.

In Asia the functional equivalent of baboons are the macaques. They are a highly successful group of geographically varying species that are found all over tropical Asia; indeed they also extend into temperate latitudes in China and Japan. Some macaques are more arboreal than others but the overall characteristic of the genus *Macaca* is their proclivity for ground-living. This adaptation is apparent rather more in their behaviour than their anatomy, and in this respect they are seen to be somewhat less specialised for ground-living than the African baboons.

Baboons and macaques are a relatively easy group to study in the field and consequently more is known about them than any other primate group. Their social structure is extremely advanced and more reminiscent of the organisation of human societies than is that of chimpanzees.

It seems that there are two groups of Old World primates whose social organisation can provide clues to the understanding of the evolution of human behaviour. For the basic aspects of our behaviour we can look to our genetic cousins the chimpanzees with whom we share an arboreal infancy; but for the more complex stages of sociality that preceded the 'cultural revolution' that led to the emergence of man, we must look to the evidence provided by the ground-living monkeys. We have to bear in mind that any behavioural similarity is not the outcome of genetic affinity but of parallelism brought about by a common environment.

In order to illustrate how environment dictates the form of social behaviour, the remainder of this chapter will be devoted to a comparison between the life styles of chimpanzees and baboons. To discuss the possible implications for the human state at this stage of our knowledge would be premature but there is no reason why each and every one of us should not draw our own conclusions. Scientists are particularly reluctant to apply evidence obtained from non-human to human primates, not because they are unimaginative, but because they are acutely aware of the dangers of drawing conclusions from inadequate evidence. It is appropriate to recall the words of the biologist Sir Peter Medawar who has described the function of scientists as '. . . building explanatory structures, *telling stories* which are scrupulously tested to see if they are stories about real life' (*The Art of the Soluble*, page 170).

COMPARISON OF BABOON AND CHIMPANZEE BEHAVIOUR

Chimpanzees are gregarious animals living in large groups of 30–40 or more in number. Sub-groups or 'bands' within the major group may be mixed, all-male or all-female (consisting of mothers and their young); mother-bands are the least mobile and male-bands the most. Members of bands may join other bands within the group, so that a certain amount of mobility, fragmentation and reunion is observed. The diet of chimpanzees is based largely on fruit, and as fruiting trees are widely dispersed throughout the forest, feeding involves a good deal of travelling and peaceable reunification of bands at preferred feeding sites. The lack of concern for territory shown by chimpanzees is not typical of other forest primates. The guenons (monkeys of the genus *Cercopithecus*) for example defend their territories quite determinedly. However, chimpanzee groups tend to avoid each other and stick to their own particular home-range though there is evidence of some interchange between groups.

Social interactions between individuals within the group are generally tolerant. Male dominance behaviour, though present, is relatively unobtrusive; this is also true of aggressive behaviour. Leadership is similarly low-geared; it exists but it isn't obvious.

Chimpanzees are extremely vocal and are quite the noisiest animals in the forest; indeed they can afford to be as they have relatively few predators. On discovering a particularly delectable food source, chimpanzees call loudly, presumably to alert other bands. They also employ an artificial means of distant communication by drumming on logs or the buttress roots of forest trees. A 'carnival' of chimpanzees, hooting, screaming and drumming, has to be heard to be believed.

Baboons are rather different. They live in troops that do not fragment into bands and change little in composition from year to year. Baboon societies tend to be 'closed' and movement of individuals from one troop to another occurs only occasionally. The animals are omnivorous and will eat almost anything in the vegetable line that is available; they are not averse to animal flesh but, at the same time, they are not hunters. Their food is everywhere, but baboons have to work harder for each mouthful than chimpanzees. Baboons have a 'territory' but do not defend it in the way that robins do for example; but they have nevertheless a strong sense of ownership. They have a home-range, a familiar stamping ground, and within the home-range are 'core areas' which contain their sleeping trees; these zones would probably be defended if challenged, but they never are. Tolerance towards other baboon troops is exhibited only at the infrequent water-holes which savanna animals are obliged to share, irrespective of territory or home-range; and sometimes, when trees in the area are scarce, they may have to share at the 'sleeping-trees'. A cadre of dominant males is pivotal to the stability and safety of the baboon troop and in-troop discipline is rigidly enforced.

Baboons communicate comparatively silently. Their vocal repertoire is extensive but is pitched in a low vocal key. Clear vision in open country replaces the need for high-frequency, long-distance calls.

The differences between chimpanzee bands and groups and baboon troops is inherent in the collective nouns used to describe their aggregations. 'Bands' implies irregulars or guerrillas trained to fight in forests or wildernesses, while 'troops' suggests disciplined ranks of soldiers disposed in strategic formations as on the plains of Waterloo. Unfortunately this analogy, though attractive, does not hold. Baboons are certainly at constant war with their environment, but chimpanzees – the flower children of the primate world – are at peace. War in the form of predation is the crucial factor dictating the social organisation of these two groups. Savanna baboons are exposed to attack by lions, leopards, cheetahs, hyenas, jackals and hunting-dogs, but have few trees to escape into and therefore have to stand and be prepared to fight. Chimpanzees and other forest primates having few predators and a multiplicity of escape routes resort to flight. Baboons meet their challenge by forming units containing a large number of adult and sub-adult males who provide the 'thin red line' of disciplined defence. Chimpanzees, on the other hand, can afford to move about in small units, often composed of females and infants completely unprotected by the males.

Thus both species show striking parallels with the ways of our own society. We show some aspects of our behaviour that seems very like that of open country primates and some like that of forest dwellers.

The 'great divide' between the apes and man was a very significant milestone. From that moment on, the hominids (Hominidae, the zoological family of man) went from strength to strength and the cultural gap between 'them' and 'us' grew even wider, which is why chimpanzees are in zoos and humans are the ones who visit them instead of the other way around.

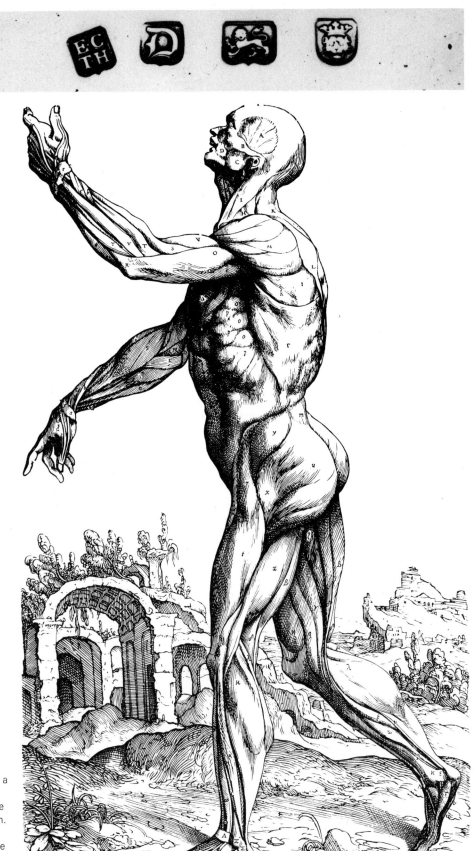

55
Hallmarks
of silver.

56
Hallmarks of
mankind. The
upright posture,
the striding gait
made possible by a
pair of strong
buttocks, a unique
human possession.
A plate from an
anatomical treatise
by Vesalius.

5

The hallmarks of mankind

Hallmarks are symbols impressed on all gold and silver objects. Their primary purpose is to provide a guarantee of the purity of the metal, but they can supply other important information such as the date and place of manufacture and the identity of the craftsman. To find the hallmark it is usually necessary to turn a piece of silver upside down and look at its bottom. If you turn man upside down and look at his bottom you will also find one of his hallmarks – a pair of fleshy pads called buttocks. Buttocks are indicators, first of all, that we are dealing with true man, but they also tell us about how he stands, walks and sits down; buttocks are one of man's hallmarks. In this chapter we shall be concerned with various hallmarks – characters that are unique to modern human beings and are therefore diagnostic of the species *Homo sapiens*.

The story of man and his ancestors is like a play in which the key character does not appear until the last scene. Yet by the time he finally makes his entrance the audience has already got a very good idea, from what has gone before, of the sort of person he is going to turn out to be.

To say that man is the key actor in the drama of primate evolution is to take a rather anthropocentric view of life. If this book was really being written by a giraffe, man might find himself allotted a very minor part in the evolutionary chronicles of the primates. It is natural that man should be self-centred in his approach to his own evolution, so we are primarily interested in those structural and functional features by which we characterise ourselves today. This being so we had better have a very clear idea of just what we are looking for in the primate fossil record. There are a number of characters that we might choose but, as our source material is limited to fossil bones and teeth, the range is naturally restricted.

The possession of speech and language is the most outstanding human hallmark of all, but unfortunately it leaves no trace in fossil bones. One can make all sorts of inferences that speech evolved at such-and-such a date but there is no scrap of direct evidence to support any such assertions. The ability to speak lies first of all in the shape and musculature of the mouth, tongue, soft palate, pharynx and larynx; and, second, in the centres of the cortex, the outer shell of the brain, which govern the muscular control of the various 'soft' parts mentioned above. Although many suggestions have been put forward, none as far as is known can help to identify the capacity for speech from fossil bones.

There are numerous cultural phenomena which we would regard as significant hallmarks, but again we cannot determine their existence from fossil bones. Behaviour does not fossilise as such, but fossil sites reveal more than the characters of the bones of their occupants and it is from the debris of the 'living floor' as it is called that a number of insights into human activity can be gained. For example a great deal of information about the diet of early man can be discovered by sifting the animal bones of the 'kitchen middens'. The importance of hunting to a community can be assessed from the type, frequency and condition of the animal bones on the floor or in the vicinity. The use of fire is clearly determinable from carbon deposits, charred bones and so on. The characteristics of stone tools are very revealing of the technological level of the late residents just as evidence of cave paintings and burials provides pointers to their cultural development. However the background to man that we are committed to investigate extends many millions of years back at a time when no living floors and no artefacts existed. So, apart from the evidence of stone or bone tools as supplements to our understanding of human manual dexterity, we shall not be leaning very heavily on the evidence of 'fossil behaviour'. What, then, are to be our criteria?

When we think about man and compare him with non-man, one of the first things that strikes us is that he stands upright and walks on two legs. However this is not nearly a precise enough definition to exclude the many non-human primates who are also capable of upright bipedalism. Nor does it exclude several non-primate mammals, for example the bears. For a more exact criterion we must draw upon our knowledge of the biomechanics of human walking.

The refinements of human walking express themselves most clearly in the foot, which being at the end of the leg and in contact with the ground should, logically, demonstrate the characteristics of the human walking pattern in the most critical manner. Comparison of the chimpanzee and human foot reveals a number of differences. Proportionately, chimpanzee toes are much longer than man's but – more significantly – the big toe in chimps is abducted, widely splayed, so that its long axis lies at an angle of 60–70 degrees to the rest of the foot. In man the big toe is adducted – it lies parallel with the other toes. This shift in the position is accompanied by a considerable enlargement of its bones and the muscles controlling its movements.

Although the big toe for some reason is a rather farcical structure, presumably because of its strong associations with gouty eighteenth-century noblemen, it is really one of man's proudest possessions, a true hallmark of his humanity. In its relative bulk and length and in its fully adducted position man's big toe is unique; no other primate has anything exactly like it. The big toe plays a critical role in human walking. When the foot strikes the ground at the beginning of a step it strikes it heel first. Because the human foot is usually turned out, it is the outer part of the heel that takes most of the force of the impact. It is easy enough to confirm this on your own bare foot because it is on the outer side of the heel that the skin is most callused.

You can double-check this fact by noticing that this is also the site of the greatest wear on the heel of your shoe.

As the step progresses your body is beginning to catch up with your foot just as a punt catches up with a punt pole (Figure 57). The result is that the point of contact of the foot with the ground extends progressively forward. By the time the body is directly above the stepping foot, the point of contact with the ground has shifted from the outer to the inner border. The weight is now being supported largely by the ball of the foot behind the big toe; the heel has already left the ground. The final phase of the step is a function of the big toe alone which is why it is so relatively huge compared with the outer four toes. The act of propulsion which drives the body forward into its next step is called the 'toe-off' (Figure 57). The whole cycle of the step, which alternately involves the right and left feet, is a rolling action in which the foot rolls forward from the point of heel-strike to the final stage of push-off from the big toe without being lifted off the ground. The nearest everyday analogy that comes to mind is the action of one of those old-fashioned rocker-like blotters.

For us to describe walking simply in terms of the foot is rather like trying to explain how a car works in terms of the wear-pattern on the tyre treads. The evidence that the wheels go round and that the car can be steered is there, but you learn nothing about how an internal combustion engine works. Human walking is an extremely complicated activity and involves a series of abstruse biomechanical concepts that go far beyond the scope of this book. Suffice it to say that the joints and muscles of the whole body are involved, not just the legs, in moving the centre of gravity of the body through space. Over many millions of years, evolutionary adaptations have been operating towards refining the human gait, turning the shambling energy-consuming waddle of a bipedal ape-like creature into the smooth undulating energy-conserving progress of a young and healthy human being. Man's walk has been described as *striding*. This

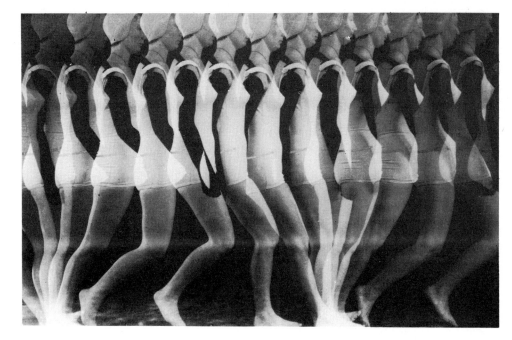

57

The action of legs and feet during striding (only right leg shown in this multi-exposure photograph, the left leg is clothed in a black stocking). Note 'Toe-off' phase second from right.

description provides the means of distinguishing between human bipedalism and the bipedalism of other primates and mammals that occasionally walk upright. Among primates, man alone strides and therefore, one argues, any fossil evidence that provides anatomical evidence for striding behaviour testifies to the human status of the specimen.

Striding may be regarded as the quintessence of human walking, a means of travelling during which the body's energy output at each step is reduced to a minimum. As such, the striding gait must have been highly adaptive for man-the-hunter who thereby was capable of following wild animals for periods of twelve hours or more with the most economical use of his physical resources. The fossil evidence of early man in Africa suggests that the evolution of a striding gait occurred well over a million years ago, a period which broadly speaking coincides with man's emergence as a hunter of big game.

One of the curious aspects of human walking is not that man walks so well but that, at times, he can make such a hash of an activity that by now should be second nature to him. This should not be taken to mean that we are all a gangling bunch of Johnnys-head-in-air who are liable to disappear down any unguarded manhole, or trip over every wrinkle in the carpet, but that as a species we seem ill-equipped to carry out the complicated manoeuvres of bipedalism without a large proportion of our number paying a penalty for the presumptuousness of our ancestors in terms of injury or disablement.

Ovid saw human uprightness as a hallmark of man's questing spirit: 'God made man erect to contemplate the heavens'. Perhaps Ovid's somewhat suicidal precept accounts for the number of individuals who actually *do* fall down open manholes.

Human walking is a very risky business, a cause of a remarkably long list of human ailments. The 'banana skins' of life await the unwary around every corner: the slick of oil, the uneven set of paving-stone, the sudden push in the back are constant threats to our shaky equilibrium. During some phases of the walking cycle the fact that man does not fall flat on his face depends on the support of an area of foot not much bigger than a postage stamp.

With training it is possible to reduce the risks of bipedalism considerably. Karl Wallenda, leader of the famous German 'high-wire' family, The Great Wallendas, recently crossed the 700-foot-deep Tallulah Gorge in Georgia, USA, walking for nearly a quarter of a mile on a tight rope. Karl Wallenda even had the nerve and energy to stop half-way across and stand on his head.

Yet Sylvia Potts, a New Zealand athletic hope, slipped and fell, whilst in the lead and within a few feet of the finishing line, during the women's 400 metres event at the Edinburgh Commonwealth Games in 1970. Training is not the whole answer.

Man is not the only primate which puts itself at hazard by depending on two limbs. Gibbons which progress by swinging from overhead branches by two arms are, as I pointed out earlier, not immune from accidents. It has been reported that thirty-three per cent of a series of 200 individuals shot in Malaysia showed evidence of healed fractures, presumably the result of locomotor accidents.

Theoretically, man would have been very much better off, as far as his physical well-being is concerned, if he had stuck to a four-legged gait. Even quadrupedalism has its own quota of disadvantages, but they are negative rather than positive. Where all four extremities are feet, given over to the function of walking, there is little opportunity for manual skills to develop. In fact many people see human bipedalism as the first step which led to the emergence of human technology, initiated by the freeing of the hands from the chores of locomotion.

Tools can only be used or made by mammals when the hands are free. The sea otter, a well-known tool-user, employs a rock to smash open abalones while floating on its back. Hooved animals do not use their hands at all for any activity other than support except perhaps for 'pawing' the ground, or in the special case of Clever Hans the German 'Wonder Horse' of counting up to any number chosen and even working out square roots! Undoubtedly it was Hans's trainer who was the clever member of this famous double-act. Carnivores, rodents and insectivores have paws – hands of a sort – and they do with them what they can but with no great success as they lack prehensility. Even quadrupedal primates, which universally possess a degree of prehensility of the hand that permits them (for example) to feed using one hand only, have a restricted manual repertoire. Their hands are essentially foot-hands. Only man has a true, fully emancipated, hand-hand.

For all its disadvantages, the assumption of upright posture was the primary adaptation that led to the emergence of the human stock. All man's characteristics – his culture, his ability to speak and use language, his technology and his large brain – might be looked upon as the consequences of standing and walking upright. Even if bipedalism is a mechanically imperfect adaptation the price we pay is, relatively speaking, a small one. After all, if the worst comes to the worst, we can still get by with inserts in our shoes, crutches and bath-chairs.

There is good reason to believe that far from being unique to man, the upright posture is a primate heirloom that has been handed down from generation to generation for millions of years. Fossil evidence from the geological time period known as the Eocene (starting some fifty-five million years ago and lasting for eighteen million years) indicates quite clearly that at this time primates were small-bodied, long-legged creatures that clung to the boles of trees and to upright branches with their bodies held vertically and their legs acutely bent at hips and knees. They are referred to as vertical clingers and leapers.

Several of their descendants are found today among the large group of relatively primitive primates known as the prosimians (lemurs, galagos, lorises, etc.). Some have persisted as vertical clingers; others, like lemurs and lorises, have evolved into quadrupeds by the relatively simple adaptive process of shortening their hindlimbs and lengthening their forelimbs. One living lemur, the ring-tail (*Lemur catta*), shows precisely the expected intermediate characters between vertical clinging and quadrupedal behaviour.

It seems highly unlikely that the human stock arose directly from a vertical clinging ancestry, but there is good reason to believe that the ancient and basic possession of truncal uprightness has dominated primate postural evolution and blazed the trail that ultimately led to the emergence of this important human hallmark. As we have said, two-footed standing and walking is not

58
One of the ever-present risks of bipedalism. Sylvia Potts, a New Zealand athlete, stumbling and falling just short of the tape during the Commonwealth Games at Edinburgh.

59
Karl Wallenda crossing the Tallulah Gorge on a tightrope.

unique to man. All higher primates (above the lemur grade of evolution) are capable of standing upright and a few actually walk upright. For example among the monkeys of the New World, the spider monkeys are great bipedalists. Even the relatively primitive squirrel monkey (*Saimiri sciureus*) given the appropriate stimulus becomes a bipedal walker. An experiment was carried out whereby the arms of an infant squirrel monkey were taped to its sides. Normally the infant would clutch its mother's fur with its hands; deprived of this safeguard it becomes quite helpless. The mother solved the problem by scooping up her infant in her arms and walking away on two legs.

The Old World monkeys rise to their hindlegs at the drop of a coconut so to speak. Indeed the need to transport food items from one place to another commonly induces bipedalism in living monkeys. Carriage of food also gets chimpanzees up on to their hindlegs. The Dutch zoologist, Adriaan Kortlandt, has filmed chimpanzees raiding native plantations in forest clearings and running back on two legs to the cover of the forest clutching as many paw-paws and sweet potatoes as they can carry. Gorillas, too, are frequent bipedalists. The males use the added height that a two-legged stance provides to make themselves more intimidating when performing the threat display that culminates classically in chest-beating.

Any or all of these activities that result in intermittent two-footedness are plausible behavioural precursors for the habitual bipedalism of man. In view of the ancientness and the universality of truncal uprightness in the primate order, it is not surprising that natural selection should have operated so strongly and so rapidly in bringing about a transformation from a quadrupedal to a bipedal way of life in our own family.

Human walking can be distinguished from all the non-human varieties by virtue of its special characteristics. For a start bipedal walking is the *habitual* gait of man; for other primate species, bipedal walking is only an occasional or part-time activity. Second, man has a special way of walking which we call *striding*. Man – and only man – strides. We can now formulate our first hallmark: *Man stands upright and when walking habitually uses a bipedal, striding gait.*

THE OPPOSABLE THUMB

The second characteristic that strikes us is the dexterity of the human hand which is infinitely capable, exquisitely delicate but, at the same time, alarmingly powerful – powerful enough to cleave a brick in half with a karate chop, or to tear a city telephone directory into two equal parts. The essential component of the human hand is its *opposable thumb*, which provides the means of grasping objects with strength (the power grip) or with delicacy (the precision grip). The opposable thumb is therefore an obvious hallmark, but unfortunately it is not unique to man. All living Old World monkeys and apes possess opposable thumbs.

Man's precision grip is much more sophisticated than any monkey's or ape's; when he places his forefinger and thumb together in a precision grip he is engaging the two most acutely sensitive areas of his whole body. The input to his brain from these small areas of skin provides the manipulative basis for the sort of skill that a watch-maker, an engraver or an assembler of micro-circuits possesses.

Firstly then we shall look at the structure of the human hand. There are four points in particular to be noted (Figure 60):

1 **Finger-tip pads.** The concentric ridges of the finger-tip pads are the contact points that subserve the sense of touch; the tiny pits that one can see with a magnifying glass which appear at regular intervals along the ridges are openings of the sweat glands which moisten the skin and thus facilitate tactile sensibility. The precise patterns of the finger-tip pads are genetically controlled; the pattern is established six months before birth; they are unique to each individual and they never change. Underlying the ridges are many free nerve endings but particularly important are the nerves which carry endorgans, minute structures shaped like a tangled skein of wool. These are called

60
Outstanding features of the human hand. For details see text.

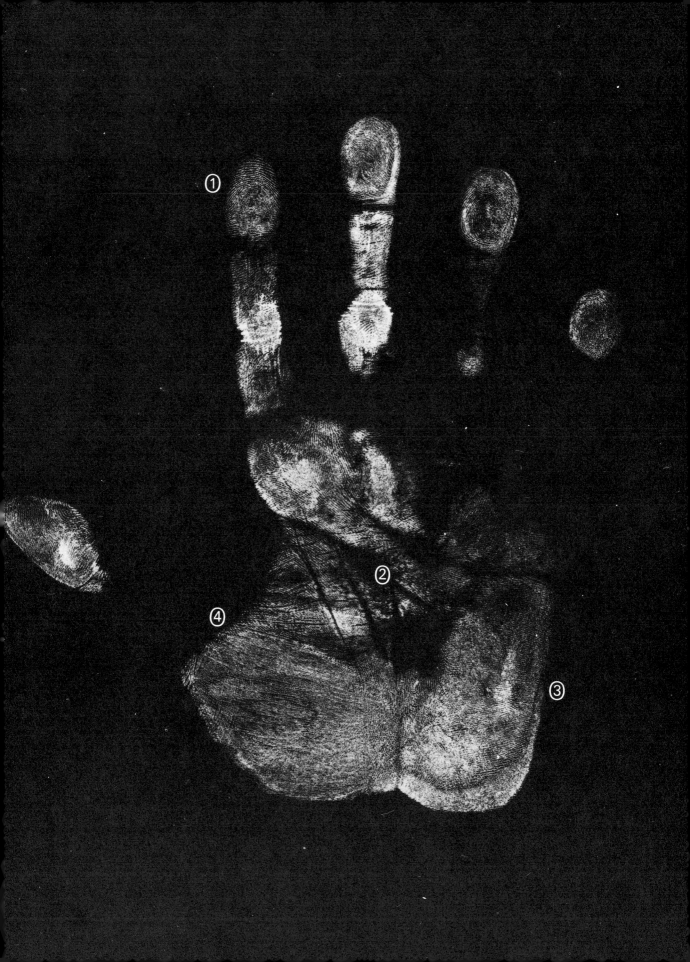

Meissner's corpuscles and provide the basis for tactile discrimination.

The finger tip itself is pulpy and adjusts itself to the shape of whatever is being gripped. The flat nail provides the backing against which pressure is exerted. All primates, with the exception of the marmosets, have flat or flattish nails instead of claws. The story goes that one day the great philosopher Plato, who was instructing a class of students which included the cynic Diogenes, defined man as a hairless creature standing on two legs. Diogenes immediately leapt up, rushed out and grabbed a chicken, killed it, plucked it, and holding it aloft shouted: 'This is Plato's man'. Plato, with infinite patience, remarked that had he, Diogenes, waited until he, Plato, had finished what he was going to say he would have heard him add 'and with broad flat nails on the ends of the digits'.

2 **Flexion creases.** Far from foretelling our emotional future, as palmists would have us believe, these are merely indicators of our pedestrian present. They are skin folds or hinges that facilitate easy bending. The 'heart line' is broken opposite the base of the index finger; this break indicates the independent movement of this digit and is a human hallmark not seen in any other primate.

3 **Hypothenar eminence.** The 'heel' of the hand of apes and man corresponds to the heel of the foot-hand of quadrupedal primates. In man the hypothenar eminence acts as a muscular buttress to reinforce the power of the hand in gripping the handles of tools and weapons.

4 **The thenar eminence.** The 'Mount of Venus' or, more prosaically, the ball of the thumb is composed of a series of small muscles which control its extremely complex movements. The special character of the main thumb joint is its power of rotation by which it can swivel inwards; the thumb joint is a highly sophisticated biological mechanism that permits the same freedom of movement as the hip or shoulder joints enjoy without their bulky musculature which would be inappropriate in the hand. Rotation permits the thumb pad to oppose one or other of the finger pads. Opposition of the thumb and index pads provides the basis of human manipulative skills. Old World monkeys and apes can oppose thumb and finger with varying degrees of success depending primarily on the relative lengths of the two digits. Effective pulp-to-pulp opposability is impossible when the thumb is disproportionately short, as it is in the great apes. A quantitative method for assessing this function is called the *opposability index* and is calculated by a simple ratio: length of thumb × 100/length of index. Man's index varies from 63–69 but has an average of 65. The average values for the apes and some of the Old World monkeys are as follows:

Apes		*Monkeys*	
Orang-utan	40	Langur	41
Chimpanzee	42	Macaque	53
Gibbon	46	Baboon	57
Gorilla	47	Mandrill	59

Man as you can see is well ahead in this respect. Apes, because they possess very long fingers and shortish thumbs, are at the bottom of the list while the long-thumbed baboons and the mandrills are nearest to man. Although most people are unaware of the functional significance of finger-thumb opposition they cannot be unfamiliar with its implications in sign language; it is the universal gesture of success – human success.

FUNCTIONS OF THE HAND

We turn now to the function of the hand and in order to discuss it, it is necessary to devise a special terminology. It is hopeless, of course, to use the ordinary language of anatomy, for the hand comprises eighteen joints, totalling altogether fifty-two different movements, and descriptions that encompassed all these would be impossibly cumbersome. Some years ago I introduced a terminology to describe movement of the hands, not in terms of individual joints but as a living whole; it is a very simple classification. In spite of the multiplicity of activities of the hand, involving countless objects of various shapes and sizes, there are only two

61
The power grip (demonstrated by strong girl Joan Rhodes, the cabaret artiste, photograph, Paul Popper)

63
The short thumb of the chimpanzee makes the precision grip a clumsy and unstable affair. Note the more effective opposition of the human fingers and thumb.

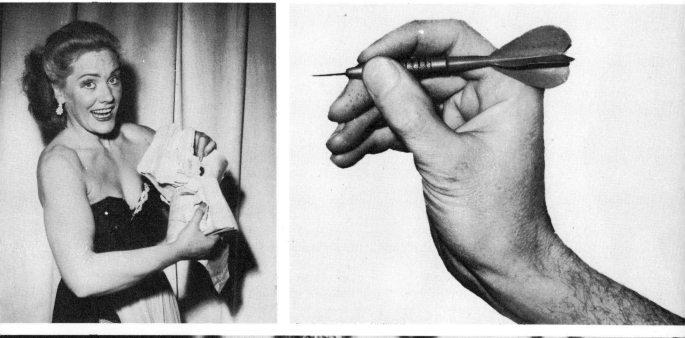

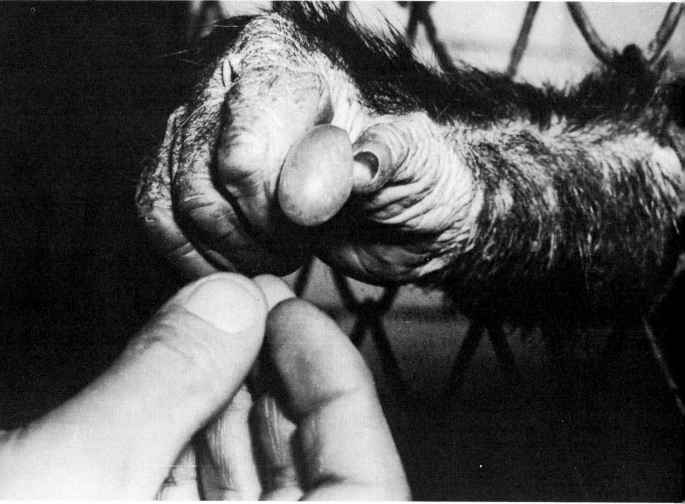

prehensile actions: these are called the precision grip and the power grip.

Which grip is used at any given moment is conditioned by the nature of the proposed activity. If precision is the more important or relevant element, then the precision grip is used. If power is the primary need and precision of secondary importance, then the power grip is employed. The shape of the object handled has little or nothing to do with the type of grip used; the sole criterion is the nature – in terms of power and precision – of the activity.

All prehensile movements of the hand consist of either a power or precision grip. This is a useful generalisation but not wholly true. There is one other grip pattern which is normal to anthropoid apes, but only occasionally used in the human repertoire. This is called the hook grip and is used as a last resort for hanging on to the edges of precipices! Although this can hardly be called a day-to-day activity, there are more prosaic functions in which the hook grip comes into its own – carrying heavy shopping baskets for instance. Apes use the hook grip all the time, not for carrying shopping baskets but in cliff-hanging activities among the branches of the trees.

The power grip was the first to appear in human evolution; the precision function is a relatively recent acquisition. This succession can be observed in the human infant whose precision grip develops some time after his power grip has become established.

We can now formulate the second hallmark. *Man possesses an opposable thumb whose length is approximately sixty-five per cent the length of his forefinger.*

THE SIZE OF THE BRAIN

The third feature that strikes us about man is that his brain is large and rounded; but of course brains do not fossilise and we can only make deductions about the brain from the study of fossil skulls. Unfortunately, apart from overall shape and size, there is no means of determining the nature of the brain by a simple examination of the skull. Size is a misleading indicator because it is so variable within a species; for example among modern human populations the brain volume ranges from 950–2000 cubic centimetres (cc). The average volume of the population is about 1350 cc. Brain size is related to body size – bigger animals have bigger brains – and, in some way that we don't fully understand, to intelligence. Nevertheless, brain size is a valuable guide to the palaeontologist attempting to follow the track of man through time. From the earliest prehuman stages to the final flowering of the human family expressed in the species *Homo sapiens*, a steady trend towards enlargement is seen. Here, then, is the basis of our third hallmark: *Man, relative to his body size, has a large rounded brain that approaches, and may exceed, 1400 cc in volume.*

THE TEETH

Finally, we notice that man possesses small, even teeth arranged in an elegant parabolic curve in his upper and lower jaw. The human teeth like those of all living primates are of four types: incisors, canines, premolars and molars. Together in both jaws they total 32, a number characteristic of all Old World monkeys and apes, but not New World monkeys (except marmosets) which have 36. Unlike apes, man's teeth are all more or less the same vertical height. The canines which form massive, elongated, projecting teeth in the apes are small, short and discreet in man and resemble the incisors in shape. Human molars have low rounded cusps in contrast to the sharp, prominent, cone-shaped cusps of apes and monkeys. The human third molar in both jaws is often small and is frequently absent, whereas in apes the third molar is often the largest of the three. There are many other differences but these few we have mentioned should be adequate for the purpose of defining the fourth hallmark: *Man's teeth are small and are arranged in the jaws in a parabolic curve, the third molar being the smallest of the series and the canines are incisor-like and non-projecting.*

EVOLUTION OF MAN

With these four hallmarks in mind (and there are many more that could have been men-

Epoch	Age	Fossil species	Gait	Hands	Brain size	Teeth	
PALAEOCENE	65	PLESIADAPIS	−	−	−	−	B
EOCENE	54	NOTHARCTUS	+	−	−	−	B
		SMILODECTES	+	−	−	−	B
OLIGOCENE	36	PROPLIOPITHECUS	○	○	○	+ +	?M
		AEGYPTOPITHECUS	○	○	−	+	?M
MIOCENE	23	PROCONSUL AFRICANUS	−	(+)	−	+	B
		DRYOPITHECUS	−	○	○	−	B
		RAMAPITHECUS	○	○	○	+(+)	?M
PLIOCENE	5	AUSTRALOPITHECUS	+ +	○	+	+ +	?M
		EAST RUDOLF	+ +(+)	○	+ +	○	?M
PLEISTOCENE	2	HOMO HABILIS	+ +(+)	+ +	+ +	+ +(+)	?M
		HOMO ERECTUS	+ + +	○	+ +(+)	+ +(+)	M
		HOMO S. NEANDER THALENSIS	+ + +	+ + +	+ + +	+ +(+)	B
		HOMO SAPIENS SAPIENS	+ + +	+ + +	+ + +	+ + +	M

Summary of the extent to which some of the best fossil primates match up to the criteria (hallmarks) laid down for modern man.

Key: Age in millions of years before present.

− No relevance.
+
+ + } Assessment of extent of relevance.
+ + +

○ No appropriate evidence available.
(+) Indicates half-a-plus.
M Main line.
B Branch line.

tioned) we ought to be able to pick up the trail of man during our journey through the past. The trip will be rather like travelling by train between two cities a thousand miles apart. Most railway systems are very complicated affairs with numerous junctions, switching points, branch lines and dead-end terminals, so we have constantly to be on our guard that we do not become shunted along unproductive lines that simply finish up at a pair of rusty buffers. There is a danger of this happening because evolution frequently involves a kind of mimicry which leads to similar characters cropping up in unrelated or distantly related forms. We have already seen for example that walking on two legs is not, uniquely, the possession of man. Within a given zoological order this phenomenon is more properly termed parallelism and the theory behind it is that, given a similar set of environmental opportunities, primates with a common ancestry will tend to evolve in a similar way. The best example of parallelism in primate evolution is that the monkeys of the New World and the Old World, which are related only through a common ancestor which lived forty-five million or so years ago, share so many physical characters that it is hard for the average observer to tell them apart.

In the earlier chapters we have followed the evolutionary line of the primates from their earliest beginnings in the Palaeocene through the Eocene, Oligocene and into the Miocene. Briefly we have considered some of the important 'divides' of primate phylogeny: the separation of the prosimians from the anthropoids, the New World from the Old World monkeys, the Old World monkeys from the apes, and finally, the Great Divide – the branching of the ape and human stocks. All that is left for us to do now is to follow the human line as far as the present day. Before doing so, however, we must take another brief look at the earlier stages of primate evolution with our four hallmarks in mind. When do they first appear?

The table provides a list of the principal fossil primates. The degree to which these fossil species display the human hallmarks of gait, hands, brain and teeth is crudely expressed by means of plus marks in columns 4–7. The last column, in the spirit of our railway journey analogy, indicates whether a fossil species is thought to be on the main line or a branch line.

The earliest primate, and one of the best known so far as abundance and skeletal completeness is concerned, is *Plesiadapis* whose remains have been discovered in North America and France. In size and shape *Plesiadapis* was something like a squirrel with a squirrel's arboreal-terrestrial habits and a squirrel's claws and limb proportions. There is really nothing about the gait, hands, brain or teeth of *Plesiadapis* which could be called an incipient hallmark of man although, dentally, it is pretty certainly a primate. It may be on the main line to man, but in view of certain rodent-like specialisations, it is probably not.

The best-known species of Eocene primates are *Notharctus* and *Smilodectes*. Both genera come from the Bridger Basin in Wyoming. In terms of anatomical detail they are the best documented of all early primates. Reconstruction of these two genera are of particular interest because they demonstrate in their limb proportions the sort of posture and locomotion that could have laid the foundations for the upright posture of man. Most authorities are prepared to classify the locomotion of these early prosimians as *vertical clinging and leaping*. This rather special type of posture and movement was first recognised in 1967 by my colleague Alan Walker and myself and has since been widely accepted as the earliest known locomotor specialisation of the primates. It is practised today by a few of the lemurs of Madagascar, the tarsier of South-east Asia and galagos of Africa. It is strictly a prosimian speciality. Its significance for human bipedality is not obvious, for after all if the Eocene fossil prosimians *Notharctus* and *Smilodectes* gave rise to anything we have with us today, which is doubtful, it would be the living prosimians which are definitely on a branch line so far as the evolution of man is concerned. However these Eocene fossil primates are representative of the stock from which recent primates have evolved and, therefore, their locomotor talents must have contributed to the genetic pattern of all living primates. The hands of *Notharctus* and *Smilodectes* were capable of prehension but the thumb contrary to the opinion of one of the greatest of American comparative anatomists, the late W. K. Gregory, was not opposable in the sense that we use the term (for man) today. The brain was probably relatively large by modern non-primate standards but the presence of a long muzzle in *Notharctus* (though slightly reduced in *Smilodectes*) suggests that *Notharctus* still depended largely on its sense of smell. The teeth were primate-like but provided no clues appropriate to our particular quest.

When we come to the Oligocene epoch two fossils, *Propliopithecus* and *Aegyptopithecus*, provide us with some clues. We know nothing of their anatomy below the level of their necks but so far as their teeth are concerned there are a few pointers. *Propliopithecus* (named because in the opinion of its first describer it was a natural antecedent of the Miocene form *Pliopithecus* which had already been diagnosed as an ancestral gibbon) possessed a dentition which had none of the specialisations of the later fossil apes called, collectively, the Dryopithecines. There were no massive canines for example. This among other features led to the tentative suggestion by Dr Elwyn Simons of Yale University, that *Propliopithecus* might represent the first appearance of the human stock in whom, of course, small canines are the rule (Figure 52). This is not, however, a popular view.

Aegyptopithecus is more ape-like and therefore less man-like. Its teeth and jaws are strongly reminiscent of later fossil members of the ape-family. Hence it rates only one 'plus' in the table.

Proconsul africanus which, anatomically, seems to be a natural descendant of *Aegyptopithecus* has been discussed in Chapter 4. It rates a plus for the general form of its teeth which could conceivably have given rise to the human dental pattern, and half-a-plus for its hands which, while possessing man-like proportions, lack the essential human trait of thumb opposability. *Proconsul africanus* is definitely on a branch line which leads to the genus *Dryopithecus* which almost certainly is a genus of apes.

We have likened this stage to a journey by train between two cities a thousand miles apart. We are running along a track at a good speed but we have had no real assurance that we are on the main line to platform one of the metropolitan terminus some 150 miles (about 14 million years) ahead of us. If we are to arrive at *St Homo* successfully we must now keep on the alert for landmarks to switch us on to the right track. No dozing off on this part of the trip.

The signal-box just coming up ahead reads *Ramapithecus*; this may be it.

It was suggested a few years ago that *Ramapithecus* belonged to the zoological family of man (Hominidae) rather than to the Pongidae or ape family, but recent studies have cast much doubt on this interpretation. Because the fossil material consists only of upper and lower jaws and teeth we cannot expect any help from evidence about the brain, the walking posture or the hands; we must depend purely on our knowledge of the teeth of apes and man. There is a saying that God made the skull and the Devil made the teeth and it is certainly true that their interpretation has caused considerable dissension, to put it mildly, amongst palaeontologists. The fossil remains of *Ramapithecus* have some man-like characters particularly in the flatness of the face and the relatively small size of the incisor and canine teeth, but they also have ape-like characters, notably the rectangular-shaped tooth arrangement, which contrast strongly with the rounded dental arcade of man. No conclusion as to whether *Ramapithecus* is ape or man can be reached at this stage so the signalman has a decision to make. Is *Ramapithecus* 'main line' or 'branch line'? He plumps for 'main line' and rather slowly and cautiously the train rattles across the points. In the table *Ramapithecus* is awarded one-and-a-half plus marks for the few man-like characters of its teeth.

The next signal-box is labelled *Australopithecus* and it comes in sight a mere 50 miles from the terminal.

Australopithecines were living in East Africa and probably elsewhere on that continent five million years ago. The descendants of this subfamily of man were flourishing in South Africa, in Tanzania at Olduvai Gorge and on the shores of Lake Rudolf in Kenya at the end of the Pliocene and during the first half of the Pleistocene; the fossil record of these times and places is very rich.

The early (Lothagam and Kanapoi) representatives of the genus *Australopithecus* do not tell us very much, but the later forms in East and South Africa show, in the characteristics of the gait, brain-size and tooth form, some of the hallmarks we are looking for. So far as bipedalism is concerned *Australopithecus* was an upright walker but whether he was able to stride as modern man does is very questionable. The pelvic bones are well known from several specimens but differ in a number of ways from the equivalent bones of modern man. No foot bones of the *Australopithecus* from South Africa are known. Thus two plus marks for gait would seem to be a fair assessment. There are too few hand bones to permit any judgement on opposability.

The brain of *Australopithecus* is man-like in its general configuration but the cranial capacity is very low. The teeth are still rather massive, but in the order of their eruption, their wear patterns and general construction they are man-like. Nevertheless it is doubtful if the particular specimens of *A. africanus* that have been discovered from South African sites are strictly 'main line'. In view of the recent but as yet unpublished discoveries (see below) by Richard Leakey at East Rudolf in Kenya, it seems possible that true man had already appeared on the scene when the Australopithecines of South Africa were still living. Clearly a form appearing later in time cannot be ancestral to an earlier one.

In 1960 at Olduvai Gorge in Tanzania parts of a skull, a lower jaw, a collarbone and hand and foot bones of a hominid were discovered by Louis and Mary Leakey. Olduvai is a twenty-five-mile-long canyon cut from the Serengeti Plain by seasonal torrents in relatively recent times. The sides of the gorge consist of a series of layers, or 'beds' as they are called. The deepest, and therefore the oldest bed, has been estimated to be at least two million years old. Until recently this discovery, which has been given the zoological name of *Homo habilis*, has been regarded by some as the earliest known representative of our own genus, *Homo*.

His way of life appears to have been that of a scavenger, a hunter of small game, and a toolmaker. *Homo habilis* was a bipedal walker and probably able to stride. His brain was still small by modern standards but bigger than that of his predecessors, and his teeth showed a slight advance in terms of size and relative proportions compared with the teeth of the Australopithecines from South Africa and Lake Rudolf. His hands were of a human type but lacked the degree of opposability possessed by modern man.

Late in 1972 a new discovery, made earlier in the year, was publicly announced by Leakey – not Louis Leakey, the great pioneer of East African palaeoanthropology and discoverer of *Homo habilis* – but his son Richard. Richard Leakey during his excavations in the highly fossiliferous areas to the east of Lake Rudolf in northern Kenya had discovered a skull which has been dated by the potassium-argon technique to be at least 2.6 million years old. This skull antedates *Homo habilis* by nearly one million years. The skull is remarkably human-like in many ways including the absence of prominent brow ridges and the possession of a cranial capacity of at least 800 cc, indicating a much larger brain than one would have expected at such an early time horizon (Figure 65). In addition to the skull, bones of the leg have been found which bear a close resemblance to modern man. To cap it all, stone tools have been found at the same time level at Koobi Fora. They are rather more advanced, technologically speaking, than those associated with *Homo habilis* at Olduvai Gorge nearly one million years later.

The importance of Richard Leakey's discovery could be very great; I say 'could be' because as I write this we are still awaiting a full scientific report on this discovery and until it becomes available it would be unwise to build any serious new theories; but one may speculate. If Richard Leakey's claims with reference to the dating, the anatomy of the skull and limb bones and the association of stone tools are fully justified, then where does man stand? We should have to suppose that man (*Homo* in zoological terms) appeared at least 2.6 million years

ago, probably more. Faced with this new information we might, for instance, have to revise our attitude towards *Homo habilis* from Olduvai Gorge, the great discovery of Richard's father, Louis Leakey. We might have to relegate *Homo habilis* to a side-branch of the hominid stock and in future *H. habilis* would probably have to be known as *Australopithecus habilis*. As one of the original describers of *Homo habilis* with Leakey and Tobias, the demotion of our one-time blue-eyed fossil boy is rather a bitter pill to swallow, but it rather looks as if such a decision may become necessary. Provided the Leakey skull from East Rudolf lives up to all that has been claimed for it, the place of *Homo habilis* in hominid phylogeny is that of a cadet branch of the family; in spite of its toolmaking proclivities and its bipedal gait it never quite made the grade and ultimately became extinct.

Up to now it has generally been conceded that after *Homo habilis* the next signal-box on the main line to man was *Homo erectus*. We used to accept that *H. erectus* was just an improved *H. habilis* with better tools and a larger brain, but this view may have to be revised in view of the East Rudolf discoveries. In the meantime *Homo habilis* rates two-and-a-half plus marks for locomotion, two plus marks for hand function, two for brain size and two-and-a-half for teeth.

Homo erectus is known from South-east Asia (Java), Asia (China), Europe (West Germany) and both North and East Africa. Early *Homo erectus* (from Java) had a bigger brain than *Homo habilis* (*H. habilis*: 656 cc; *H. erectus*: 935 cc) and the later manifestations of this species, from Peking for example, showed a maximum brain-size of 1225 cc. In spite of this large brain volume (two-and-a-half plus marks), *H. erectus* possesses a skull of primitive and easily recognisable shape. His gait is assumed to have been both bipedal and striding (three plus marks). The form of his hands is unknown, so the only guide to the extent of his dexterity is the complexity of tools that he made. These fall broadly into the class of 'power tools', stone artefacts of simple construction designed for relatively crude purposes such as killing and

65
The East Rudolf Skull believed to be at least $2\frac{1}{2}$ million years old.

skinning animals, cutting wood and pounding vegetable products. It has been shown by experiment that tools of such simple form could have been constructed and used in the absence of an advanced form of precision grip.

Perhaps it was an increase in brain-size that stimulated the evolutionary improvement of the hand, or perhaps it was the other way about. Anyhow it seems probable that the complexity of the brain, the precision capabilities of the hand and the evolution of 'precision tools' like slender blades, stone chisels and bone needles, were closely interlinked.

Exactly where and when *Homo erectus* evolved into *Homo sapiens* is not known. It may have happened in different parts of the world at different times and in different ways and there is no saying in which geographic population of early men the species first emerged.

With the evolution of *Homo sapiens*, which is dated somewhere between 250,000–400,000 years ago, our railway journey is almost at an end. There is only one further switch point left that we have to be wary of.

Neanderthal Man (*Homo sapiens neanderthalensis*), the quintessence of the ancient caveman image, is undoubtedly on a branch line. To speed along this particular section of track leads simply to the end of the line and the corroded buffers of racial extinction, which is precisely what happened to poor old Neanderthal Man.

We are now entering the suburbs of the metropolis. As we approach the main-line terminal, most of us can begin to put on our overcoats and lift our cases down from the rack. The engine-driver has read the signals correctly, the signalman has done his job and our worries are virtually over – over for some but not for all. The complexities of the suburban system are still to be negotiated and for some anthropologists this is the most exciting part of the trip. Their sphere of interest lies in the growth of agriculture and animal husbandry, of citizenship, of social and political systems, of the spread of populations and the intermingling of genes that are leading slowly but inexorably towards the eventual unification of mankind in a single biological and cultural entity.

Postscript

The title of the sixth and last lecture for young people in the series *Monkeys without Tails* at the Royal Institution over the Christmas holidays 1970–1 was *WHAT IS THE IDEA OF SHOOTING AT US?* The question, which ostensibly emanated from the simple and unsophisticated mind of our resident giraffe 'Geronimo', speaking for all wildlife, is both plaintive and puzzled – and rightly so. So we ask ourselves the same question, what *is* the idea?

There are five principal reasons for killing animals – self-defence, food, medical research, commercial profit and fun, and only the first two (and, in certain circumstances, the third) are in any way defensible. When man needed to kill in order to survive, the use of weapons was entirely acceptable. Unlike animal predators with their claws and teeth, man has no natural weapons for defence or attack; even his much vaunted upright posture renders him, by virtue of his exposed abdomen, exceptionally vulnerable to serious injury.

I am preaching a counsel of perfection. Few of us who dislike killing animals would go so far as to become vegetarians. However we willingly refuse to wear furs, decorate our homes with animal skins or buy foods like pâté de foie gras or white veal that involve an objectionable exploitation of the animals concerned. Far from being ashamed of our inconsistencies – the outcome of the essential duality of our natures which conveniently distinguish between animals killed for food and animals killed for fun, we should be delighted that our thinking is as advanced as it is. Ours is not the most perfect of perfect worlds. We live in a universe where even the tiniest step in the right direction should be counted as an amazing victory in the cause of moral grace. Conservation and ecology are the in-words of the moment. The fact that they are fashionable and trendy should not blind us to the essential value of the concepts they stand for.

It is inevitable that crusades such as these (like fashion, the pop scene, Jesus movements and women's liberation) attract the extremists who by their self-indulgent extravagances tend to do their cause more harm than good. Nevertheless such people, unwittingly perhaps, serve the greater purpose. When the carnival is over and the kissing has to stop, when the communication media grow tired of the novelty, and the serious people are left to pick up the badges, streamers and favours of a forgotten cause, it will be found that the battle is far from being lost; on the contrary, in some freakish way, it is half won. All publicity, as someone once said, is good publicity.

Our giraffe and his friends need no longer worry about being shot, stuffed and mounted for the edification of the masses. Principally through television, Geronimo and his kind are as familiar to the average citizen as a cow or a rabbit. From now on if Geronimo feels a brief pricking pain in his backside it will not be from a bullet fired with questionable motives but from a tranquilliser dart fired with the best of intentions. What he does have to worry about, however, is not man's wish to destroy him directly but man's blind refusal to recognise that by manipulating natural habitats by mining, logging and agricultural expansion, he is destroying wildlife as surely as by putting a bullet through the heart.

Simply expressed, our crusading fervour should now be directed towards the preservation of animal habitats. Charles Darwin in his wisdom asserted that given a viable habitat, animals will adapt and survive. The laws of nature are there; all we need is a measure of concern to see that such natural laws are allowed to operate without unnatural interference.

RECOMMENDED FURTHER READING

The life of primates, by A. H. Schultz. Weidenfeld and Nicolson, 1969.
An authoritative book written in an easy-to-read style with plenty of illustrations dealing with all aspects of primate biology.

The roots of mankind, by John Napier. G. Allen and Unwin, cased and paperback, 1971.
An account of the anatomy and evolution of the primates and man with special reference to their environments.

The apes, by Vernon Reynolds. Cassell, 1968.
A popular but detailed account of the discovery, the mythology and the behaviour of the apes – the gorilla, the chimpanzee, the orang-utan, and the gibbon.

The mountain gorilla, by G. B. Schaller. University of Chicago Press, 1963.
Dealing with the life-style of these wonderful animals as seen by a zoologist who lived among them.

The year of the gorilla, by G. B. Schaller. University of Chicago Press, 1964; Collins, 1965; Penguin Books, 1967.
A popular version of Schaller's more technical work, quoted above.

Monkeys and apes, by Prue Napier. Hamlyn All-colour paperbacks, 1971.
An inexpensive, accurate paperback which summarises all that the average reader would want to know about primates, plus a little more!

Fossil man, by M. H. Day. Hamlyn All-colour paperbacks, 1969.
An authoritative and inexpensive introduction to the fascinating subject of digging-up man's past.

In the shadow of man, by Jane van Lawick-Goodall. Collins, 1971.
A well-told and highly informative account of Jane van Lawick-Goodall's long, first-hand experience with chimpanzees at the Gombe Stream Reserve in Tanzania.

The evolution of man, by David Pilbeam. Thames and Hudson, cased and paperback, 1970.
A well-illustrated, semi-popular text on human origins.

The naked ape, by Desmond Morris. Cape, 1967; Corgi, 1969.
This well-known book contains much for the interested reader to consider and digest on the subject of man's behavioural debt to his past.

Index

DATE			
APR 1 3 1990			
OCT 2 7 1995			